Barron's Art Handbooks

TEXTURES

Barron's Art Handbooks

TEXTURES

BARRON'S

CONTENTS

ACADEMIC PAINTING

Painting displays reality to the viewer in the form of a complete illusion;
the different objects represented in the painting actually lack the volume they appear to have.
And it is not only the volume that is represented but also the natural appearance of the object
used as a model. Painting, throughout history, has traditionally attempted to imitate reality,
evolving as a result of the techniques developed by different painters. Before the camera
came into existence it was important for the artist to be able to recreate the texture
of any part of the model. This tradition of academic painting has survived
until the present day.

Graphical Information

When painting was the only method of preserving an image, artists needed to invent pictorial techniques that could be used to imitate all the features of the model on a flat surface.

We have few remains of painting from classical cultures (Egypt, Greece, Rome, etc.), although frescos and mosaics suggest that they succeeded in developing highly advanced techniques for the representation of objects, persons, and animals.

After the fall of the Roman Empire, the Middle Ages avoided all representations of nature, and it was not until the Renaissance period, when learning and pictorial investigation were again encouraged by artists such as the German-born Dürer, and painting turned to the study of birds, animals, etc., that artists again aimed to represent reality as it is.

Albrecht Dürer, Hare, (1507) watercolor. Albertina Library, Vienna.

Tradition in the Twentieth Century

Painting in the official schools was based on the recovery of classical forms, in such a way that the different canons and methods of representing the subject were adapted to the aesthetics and tastes of each period.

Academicism has left its mark on all movements, whether they were for or against it.

Since Impressionism and the invention of the camera during the latter decades of the nineteenth century, painting has branched out into many completely different styles; some of these aim to represent reality while others negate any figurative point of reference.

Salvador Dalí (1904–1989), Bread Basket. *Dalí Museum, Figueras.*

Merit and Craftsmanship of Academic Painting

Academic painting focused on specific forms and themes, and any divergence from these canons was rejected by the official critics.

With the arrival of avant-garde movements, two parallel artistic movements were established: one developed its aesthetics within the limits of academic taste while the other encouraged the painter's ability to represent reality, a skill that was held in high esteem by bourgeois society and the official critique. The avant-garde movements did away with these considerations in the recognition of art.

Illusion and Perfection

The French Academy encouraged excellence among its artists by means of prizes and awards, leading painters such as David or Ingres to develop highly refined pictorial techniques.

The paintings of French Neo-Classicism were enriched with surfaces that went beyond reality, creating highly detailed finishes for clothes, reflections, brocades, ceramics and even skin.

Jean-Baptiste Simeon Chardin (1699–1779), The attributes of Art and its Merits. *Hermitage Museum, Saint Petersburg.*

Ingres, Madame Moitessier (1833–1856), oil on canvas. National Gallery, London.

Challenging Reality

Although the French Academy was a major influence in the work of these artists, many of them, despite having been taught within the strict limits of the institution, were eventually to disagree with the themes proposed by the Academy which all too often demanded the depiction of mythical or national heroes.

As a consequence, many excellent painters placed their skills at the disposal of what was to become Realism.

MORE INFORMATION

• Texture and realist painting **p. 10**

TEXTURE AND MATTER PAINTING

With the evolution of artistic avant-garde movements, painters developed a new concept of reality, one that freed them from confining themselves to constructing the details of the model. Textures came to be seen as the effects of light perceived by the viewer directly from the painting, rather than an imitation of reality. This led to a considerable evolution of painting techniques in both academic painting and avant-garde works, creating results that were far removed from the flat, varnished surfaces the Academy demanded.

Rembrandt van Rijn (1606–1669), Self Portrait. Mauritshuis, The Hague. Rembrandt's approach to pictorial treatment had an influence on avant-garde movements.

Jacques-Louis David (1748–1825), The Death of Marat (1793), oil on canvas. Brussels Museum. The surface of the painting shows no alteration of any kind, although the textures of the model have been expertly represented.

The Texture of Academic Painting

Academic painting in the nineteenth century attempted to imitate reality following rigid guidelines that precluded the slightest variation. The paintings belonging to this period reveal a high level of technical perfection, for all the textures are obtained using tonal values to model the objects and the paintings are meticulously finished, eliminating any traces of brushwork and leaving a totally flat surface with no textural alterations.

1874: The Arrival of Impressionism

With the development of realist painting by Courbet (1819–1877), artists began to paint outdoors; the only variation from academic painting was the subject, yet the treatment of textures and the surface of the painting remained similar. With the arrival of Impressionist painting on the artistic scene, the surface of the painting changes radically. Forerunners of matter painting existed, such as Velazquez, Goya, Rembrandt, Turner, and Constable among others, however these great masters were respected and admired as individual phenomena. The Impressionist movement was characterized by strong, free brushwork and large quantities of paint that were left intact after their application.

Different Forms of Expressionism at the End of the Twentieth Century

Throughout the twentieth century, different pictorial styles have resulted in the development of many different plastic techniques that partially coincide; some have survived while others have faded quickly. At the end of the century artistic attitudes embrace all possible forms of representation, from the most academic to the most expressionist.

MORE INFORMATION
· Academic painting p. 6

Pierre-Auguste Renoir (1841–1919), Le Moulin de la Galette, Musée d'Orsay, Paris. Impressionism replaced value painting with direct, striking painting.

Abstract Expressionism and High Relief

The painting of the early avant-garde movements that developed during the twentieth century aimed at painting different interpretations of reality and practicing different styles until Vassily Kandinsky (1866–1944) painted his first *Abstract Watercolor*. This work of pure abstraction was the precursor to Abstract Expressionism, developed by Jackson Pollock, and the starting point from which textures in painting were to have an intrinsic value, independent of their function in figurative representation.

The Development of Matter Painting

Impressionism opened the doors to artistic ideas that were to lead to the new aesthetic movements that developed throughout the twentieth century. Vincent van Gogh (1853–1890), one of the major Post-Impressionists, gave rise to Expressionism; Paul Gauguin (1848–1903) adopted the ideas of the Nabis and Fauvists; all these avant-garde movements, now far removed from academic painting, adopted a set of new values, concerning both the textures of the objects painted and the themes themselves. One of the most eclectic painters of the early avant-garde movements was Henri Matisse.

Henri Matisse (1869–1954), Portrait of Derain (1905). Tate Gallery, London. Notice the salient brushwork of the painting.

Ramón de Jesus (1965–   ). When He Looks out of the Window, I Believe (1990), acrylic on canvas. Oda Gallery, Barcelona. The gesture and the substance, two key characteristics of the Informalism of the later decades of the twentieth century.

TEXTURE AND REALIST PAINTING

Realist, figurative painting is that which represents the forms of the model as faithfully
as possible, creating an illusory effect, or something as similar to reality as possible.
Although the figurative painter does not always confine himself to showing reality
in the strictest terms, a good artist does not copy the model but uses it for
his own personal interpretation. Forms and textures are always
interpreted according to different artistic criteria.

The Aim of the Realist Painter

Realist-figurative representation uses technique to display the object chosen as the model for the painting as faithfully as possible. Despite this common purpose in figurative painting, the subject can be expressed as the artist perceives it and its texture. For example, creating the texture of a model in chiaroscuro was developed by Caravaggio during a movement called Tenebrism, whereas other artists developed their own techniques for representing the different textures of the model until they reached a stage, developed by the Impressionists and Post-Impressionists, in which the texture of the object is conveyed only through colors.

Caravaggio (1573–1610),
David, the Conqueror of Goliath.
Prado Museum, Madrid.

The Study of Light when Modeling Forms

The texture of different objects can be interpreted in the way the light shapes and models the forms. The shadows created by the objects are sharper or softer depending on the light source, and on whether the light is natural or artificial.

The texture of the different objects represented in a realist-figurative painting is in complete accordance with the light that bathes the subject, and the different surfaces may take on a warm or cool tonality depending on the color of the light.

*Jean Baptiste Camille Corot
(1796–1875),* Young Woman
next to a Well. *Köller-Müller
Museum, Otterlo. The texture
of each of the different
elements of the painting is
shrouded in a cool light.*

From Detail to Synthesis

During the final decades of the nineteenth century, France saw a radical change in the creation of texture in painting.

In a striking departure from previous concepts, color started being used not only to represent the actual color of the objects but also to create planes which, in the eye of the observer, appear to blend into each other, creating highly luminous textures.

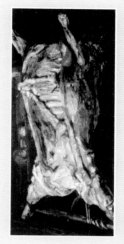

An Example of Mastery

The way in which textures in contemporary figurative painting are applied are based on plain ink-works, a genre that appears not to belong to the avant-garde movements of the twentieth century. This, however, is not entirely accurate as there have been great artists who, due to their mastery and skill, were ahead of their time in both the treatment of their subjects and textures. In his painting *The Skinned Ox*, Rembrandt seems to anticipate many of today's expressionistic movements, captured in the strong, gestural brushstrokes of this piece.

Rembrandt (1606–1669).
The Skinned Ox.
Musée du Louvre, Paris.

Vincent van Gogh (1853–1890), Fourteenth July in Paris (detail). Jaggli Hahnloser Collection, Winterthur. An accurate and concise depiction of fireworks by Van Gogh.

Textures in Realist Painting

Drawing on the ideas of Colorism, in which colors are used as planes and not values, many modern artists use pure colors to create works of great plasticity by separating the different tones with planes of pure color. This is the case with Edward Hopper and so many other painters who, though not Impressionists, recovered part of the color theory of the early avant-garde movements to create textures without needing to resort to the techniques used by Neo-Classical painters.

Edward Hopper (1882–1967), The Barber's Shop. New York University. The different planes in the painting are established by precise, well-defined colors.

MORE INFORMATION

· Academic painting **p. 6**

IMITATING DIFFERENT TYPES OF SURFACE

Creating different types of surface in painting is one of the main ambitions for most artists who practice the technique of *trompe l'oeil*. This French term, meaning literally "trick the eye," refers to the creation of illusory effects that many artists are extremely successful at. Its originality lies in its ability to deceive us: a broken pane of glass, a worm-eaten frame, metallic surfaces, marbles, and even flies that appear to be on the painting; all these effects can be painted without the observer perceiving the deception.

The Earliest Examples

From the beginning of pictorial representation, artists have attempted to reproduce reality, or a part of it, in their works. The observer will thus discover a series of features, such as a *texture* that appears to be almost real, that will serve as an introduction to the painting.

Roman art (under the influence of Greek art) contains many murals that imitate the texture of stone or plants. With the fading away of the Middle Ages and during the Gothic period, artists again began to imitate the texture of cloth and marble in their paintings.

Page from the Book of Hours *by María de Borgoña (before 1477). Painting on parchment.*

François Ferrière (1752–1839), Poetry: Amorcillos *(oil on wood). There is no relief here whatsoever; all is painted.*

The Rise of Illusory Art

After the Renaissance period, painting incorporated the techniques for perspective developed by the sculptor and architect Brunelleschi (1377–1446); these acted as the basis for trompe l'oeil. Painters took advantage of this knowledge of perspective to create spaces within planes and to represent surfaces that looked

Baldassare Peruzzi (1481–1536). *Mural fresco from the Hall of Perspective (about 1515), Villa Farnesina, Rome.*

as real as possible. This knowledge also allowed artists such as Baldassare Peruzzi to paint walls that acted as extensions to a room, imitating marbles, sculptures, mosaics ...

Technique and Hyper-Realism

The development of pictorial techniques, particularly in oil painting, enabled artists to paint all kinds of textures with incredible realism. Such high degrees of perfection were achieved that the texture of paper, wood, or marble, painted by a master of technique, is hardly distinguishable from the real thing. This mastery at painting the different objects in a subject has produced works that go beyond reality. Although Hyper-Realism appeared as an artistic movement in the United States towards the middle of the twentieth century, its origins can be traced back to the original trompe l'oeil, such as the kind developed by Francis van Myerop.

Francis van Myerop (1640–1690). *Still life with Birds (1670), oil on wood.*

Trompe l'Oeil and Contemporary Art

Painting highly realistic textures continued to be developed in contemporary art to achieve effects particular to the period. Thus, it is hardly surprising that with a tradition such as that anticipated by the great masters of painting, from the Renaissance to the present day, artists such as Pistoletto (1933) or Carlos Pazos (1949) should have made use of trompe l'oeil on more than one occasion.

Pistoletto.
Patient and Nurse *(1967), cut and painted photograph, adhered to polished steel.*

Skill and Practice

Creating Hyper-Realistic textures requires great technical knowledge of the medium being used, as well as continuous practice. Using an airbrush, such as Hideaki Kodama, Hyper-Realist artists obtain textures that go far beyond the pure and sharp qualities of a photographic image.

Hideaki Kodama, Bugatti. *Notice how subtly the shiny metal surfaces have been painted.*

MORE INFORMATION

- Academic painting p. 6
- Texture and realist painting p. 10

THE MEDIUM

MATTER PAINTING

Painting need not always result in a flat, smooth surface; pictorial media such as oil and acrylic allow large amounts of matter to be used so that the textures not only appear real as painting but also create volume and form, increasing the plastic effect and giving rise to another form of pictorial perception.

Notice the strong textural effect left by van Gogh's brush. Red Vineyards in Arles. *Pushkin Museum of Fine Arts, Moscow.*

A Long Process

In order to obtain high quality results when creating different kinds of textures, the most important factors are practice and the observation of a variety of objects using different types of lighting. Red velvet does not have the same appearance under a certain type of lighting as a simple red piece of cloth. The highlights and the folds define its texture. In order to develop painting of high quality, a great deal of time and practice is required.

The Earliest Appearances

Velazquez (1599–1660), Rembrandt (1606–1669) and Goya (1746–1828) anticipated the Impressionist theories, though at the same time their paintings revealed a strong tendency towards matter painting. Flowing, well-loaded brushstrokes leave a mark on the canvas without spreading or blending the colors. With Impressionism, Colorists developed a new treatment of matter textures, reaching a climax with van Gogh's (1853–1890) gestural painting in which the brushstroke sculpts the form to produce a strong expressionistic effect.

Textures and Styles

Within the possibilities of matter painting for creating textures,

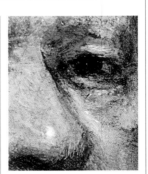

Rembrandt painted works of a highly textural nature. Self-Portrait as an Elderly Man *(detail). National Gallery, London.*

we can distinguish between two highly different tendencies: one has its origin in the Impressionist and Post-Impressionist movements; the other tendency includes the different possibilities of Expressionist and Informalist painting in which the texture of the painting acquires its own aesthetic value.

Matter Painting and Collage

Collage is an excellent basis for developing many different textures in matter painting, as the pictorial medium can be

MORE INFORMATION
· Pictorial media: Acrylic **p. 24**

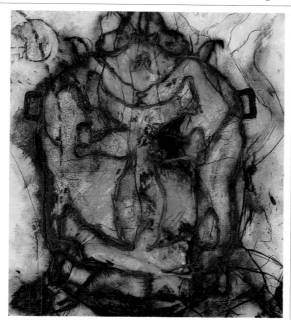

Quico Estivill. Dues Pells Ques Es Pregunten *(1992), combined techniques on canvas. Private Collection, Barcelona. The texture forms part of the expression of the painting.*

Isidre Nonell (1873–1911). Still Life. *Museum of Modern Art, Barcelona. The texture is the result of the optical effect of the painting.*

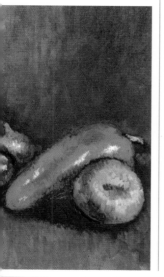

Antoni Clavé (1913). Guerrer au Masque *(1981). Painting on marouflage on canvas. Private Collection. A good example of the application of textures in Informalist art.*

alternated with other media to produce an unlimited number of textural effects. Since Informalism there have been many artists who have used collage or mixed techniques to develop their work, although on many occasions it resembles *Ready Made* (an artistic movement launched by Marcel Duchamp in 1912 that advocates using lost objects as a key element in the painted work) more than painting.

Preparing Additives

This is the work the artist carries out in his studio, altering the medium with additives to obtain the desired texture or effect in the painting. This method requires the artist to practice and experiment with pictorial techniques so as to obtain the maximum potential it offers in terms of representing different textures. Professional artists, of course, know how to create any kind of texture, this however, is not used as a mere addition to the work but is an essential part of it.

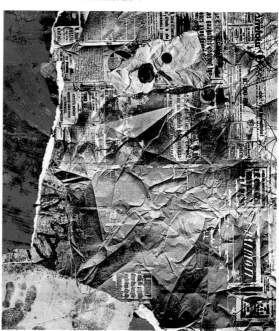

DRAWING:
TEXTURE IN CHIAROSCURO

Drawing is the main vehicle for all artists, as it is the drawing that provides
the structure of the forms. It also enables the painter to successfully represent all kinds
of textures using tones. This makes drawing particularly useful for preliminary work and
notes. Many drawing media exist; we consider any medium that can
produce a tonal gradation and hatchwork a drawing medium.

Drawing Media

Paper is the usual surface for drawing media. These can be dry: white chalk (A), charcoal (B), conté square stick (C), conté pencil (D), or in round sticks (E), charcoal pencils (F) or in sticks (G), graphite square stick (H), or round graphite stick (I), sepia sticks (J), white pastel stick (K), or metal lead (L), although this is no longer used. All these dry media can create a gradation of tones depending on how hard the medium is, and all can be gradated by blending or by drawing a set of lines that let the color of the paper show through. This means that any texture can be represented using lines or gradations of a single tone.

Different drawing media.

Peter-Paul Rubens (1577–1640). Portrait of Elizabeth Brandt, lead pencil, sanguine and white chalk. Notice how the artist has represented the texture of the skin.

MORE INFORMATION
• Academic painting **p. 6**
• Texture and realist painting **p. 10**
• Drawing textures **p. 60**

L

B

A

C

D

E

F

G

H

I

J

K

8 B	
7 B	
6 B	
5 B	
4 B	
3 B	
2 B	
B	
HB	
F	
H	
2 H	
3 H	
4 H	
5 H	
6 H	
7 H	
8 H	
9 H	

Different lead pencils and their darkest tone.

Pencils: Tonal Scales

Lead pencils come in different degrees of hardness and are grouped into ranges. They can be blended to create any kind of texture. Pressing down as hard as possible on a soft lead pencil (those in range B), will produce a gray that becomes virtually black the higher the number is; in other words, an 8B pencil will produce a darker tone that a 7B, and so on. H-range pencils are darker, the lower their numeration. B pencils are ideal for artistic drawing as they can be easily erased and produce a wide range of tones.

The Combination of Charcoal and Sanguine

Charcoal, either in stick or pencil form, can be combined with sanguine to obtain lined textures or blended textures with a warm hue.

The potential for texture in charcoal and sanguine drawing is as varied as it is rich, to such an extent that lines can be alternated with the color blended

on the paper. In addition, since they are not highly stable media, they can easily be altered, and whites can be created using the eraser, allowing the artist to create strong contrasts of light that serve to model the forms in the style of chiaroscuro.

Light and Texture with White

Drawings in pencil, charcoal, or sanguine always play with the contrast created in relation to the ground of the paper, that is, the lightest or most luminous tone is the color of the paper itself. However, if you decide to use paper that is not white, you can use its color to reinforce the highlights and create volume in chiaroscuro.

The white most commonly used in "dry" drawing is white chalk, although other techniques also use white crayon and, naturally, pastel.

An Example to Follow

The best method for understanding drawing and the representation of different textures is by observing how other artists have solved the problems that textures can present. Each surface has its own texture; notice how Thomas Gainsborough achieved the texture of this satin dress by using colored paper, charcoal, Sienna and white colored chalk.

Thomas Gainsborough (1727–1788). Study of a Woman. British Museum, London.

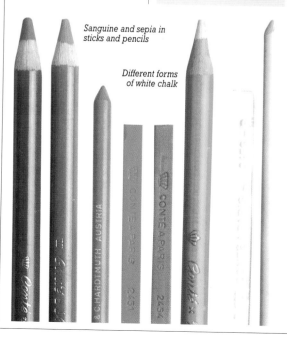

Sanguine and sepia in sticks and pencils

Different forms of white chalk

DRAWING:
TECHNIQUES USING INK; LINES

One of the most commonly used media in "wet" drawing is ink. It is important
that the learner familiarize himself with all the effects that are possible with ink.
A good starting point is to experiment with different drawing instruments to
create varied effects, and learn how to combine different techniques
to create the illusion of stone, wood, cloth, etc.

Lines

Ink can be applied with a
brush or any other instrument
that is sufficiently rigid so as to
draw an even, steady line.

A line drawn with a brush usually varies in width depending on
the amount of ink used and the
pressure applied. The line can
taper out towards the end until it
is no thicker than the very tip of
the brush. This is an excellent
rea-son for acquiring good quality brushes.

Lines drawn with a pen, of
course, are always of the same
width. The textures you can obtain with lines drawn in ink vary
according to the contrast they
create with the paper.

Lines drawn with a nib can
vary in width depending on the
pressure you apply.

Lines drawn with a brush.

*Lines drawn with
a pen.*

*Three ways of obtaining different
results when applying ink
with a brush.*

Drawing: Textures in Chiaroscuro
Drawing: Techniques Using Ink; Lines
Drawing: Techniques Using Ink; Hatching . . .

19

Painting with a Brush

Painted patches of ink, in other words, those applied with a brush, are fundamental to creating textured surfaces.

When one area overlaps with another, the resulting layer is darker due to the addition of the two tones.

If freshly applied ink is soaked up with blotting paper, a gray area is left that can then be used to obtain medium tones.

Using a brush that is practically dry, you can create a striking, "broken" effect.

Absorbing Ink

Ink can be used in drawing to obtain a variety of results that can be employed in different situations.

A blob of ink can be absorbed with a brush, a rag, or with paper towels. A mass of color can be blotted out to create a wide range of effects. For example, paper towels, which are highly absorbent, can be used to lighten the tone of large areas. A rag stained with ink can be used on the paper like a stamp to create an even texture.

The palest area was absorbed with paper towels.

A cotton rag stained with ink can be used to create textures like this.

The Texture of a Wash

Washes of ink are different gradations of the tone of the ink obtained by thinning it down. Any kind of ink may be used, though there are some types that produce a warm effect when gradated, allowing the artist to create a wide range of tones for different textures; an example is sienna ink. Other types of ink, such as the blue ink used in ball-point pens, serve to create a beautiful iridescent effect when applied in large amounts over a certain area.

Gray Textures

With ink, you can obtain grays that range from jet black to white, making it possible to imitate any kind of texture. Ink is one of the best drawing media for representing all kinds of objects; however, it requires a certain degree of mastery because, unlike pencil drawings, no correcting is possible. This means that the artist has to develop his range of grays starting with the white of the paper, and leaving untouched the areas that are intended to be the lightest in the finished drawing.

Giambattista Tiepolo (1696–1770). Adoration of the Virgin and Child, pen and sienna wash. National Gallery of Art, Washington. The texture of the clothes depends on how the tones are gradated in accordance with the background.

MORE INFORMATION

• Drawing: Texture in chiaroscuro **p. 16**
• Drawing: Techniques using ink; hatching drawn with a pen and reed **p. 20**
• Drawing textures **p. 60**

DRAWING:
TECHNIQUES USING INK, HATCHING
DRAWN WITH A PEN AND REED.

Ink drawing makes use of a variety of instruments, the brush being the most common. However, the ideal drawing instrument for creating textures is undoubtedly the pen, either metal or reed, for it allows the artist to create varying lines, and therefore textures, depending on the kind of pen that is used.

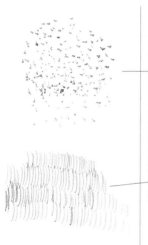

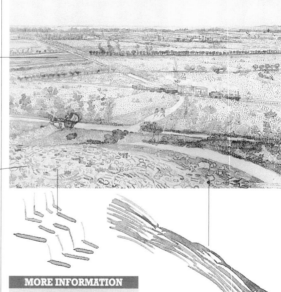

Vincent van Gogh (1853–1890), La Crau from Montmajour. British Museum, London. Notice the different textures and detail.

Crosshatching and Texture

Crosshatching is a gray area obtained by drawing a series of crisscross lines. This optical effect can be used to establish the light and dark areas of a drawing as well as its texture.

Hatching is a series of parallel lines, and crosshatching is when these lines cross over each other to form different shades of gray, depending on how dense the hatching is.

These different methods of establishing grays are used to obtain a variety of textures.

MORE INFORMATION

- Drawing: Textures in chiaroscuro **p. 16**
- Drawing: Techniques using ink; lines **p. 18**
- Drawing textures **p. 60**

Each form is a set of closely drawn lines

Law of Approximation

The Gestalt movement or school that carried out an analysis of forms, put forward a series of psychological answers to explain the visual phenomena that enabled the observer to understand iconographic language. One of these laws explains exactly why as a result of their proximity, the human eye perceives a series of points or lines as a mass of texture.

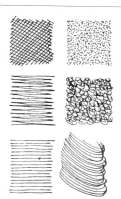

By using different types of hatching it is possible to obtain gradations and all kinds of texture.

Different types of line drawn with a nib.

Nibs and Textures

Nibs and reeds produce very different results when used for hatching and drawing lines. Nibs draw clean, uniform lines whereas reeds, which are much more fibrous, leave rough edges. Reeds are good for creating a fuzzier, warmer line that can be used for medium, gradated grays, similar to those obtained with a pencil.

Optical Effects

Textures created with a nib or a reed are the result of an optical effect; lines are repeated in a uniform pattern in a certain area of the painting and the eye perceives them as a textured mass. Different patterns can be used for different areas so that the observer associates the type of line with the area it covers, distinguishing it from other parts of the drawing covered with different types of hatching. By using different forms of optical expression, the light areas are distinguished from the planes that form the object.

The main difference between the textures produced by the nib and those produced by the reed lies in the sharpness of the line.

From Intense Black to Gray

Nibs can be used to create a vast range of grays that result in different effects.

Closely drawn lines will produce an almost black color, while more separated lines can create the texture of medium tones.

When a nib is used to paint only the black areas of the model, it is called block style; this is a type of drawing in which medium tones do not appear.

Gradations created with a nib can resemble any type of texture; the highlights on skin, backlighting, etc.

Ink, pens, and bamboo reeds.

PICTORIAL MEDIA: OIL

Different pictorial procedures provide different answers to the creation of textures.
Oil is a dense medium that conserves its appearance after drying: it does not shrink in
volume, and can therefore be used to paint a surface on which the brushwork is intended
to be noticed. Among the most interesting features of this pictorial medium are its slow
drying and the possibilities it lends the artist for blending colors. Oil is an ideal medium
for producing both realistic as well as thick textures in painting.

The Medium and Its Texture

Oil is a pictorial medium with its own particular kind of texture. Because it's a dense medium, it is ideal for all kinds of blending on the canvas, and for creating the details of the painted object.

The texture of different objects painted in oil can be created using a wet ground over which different colors are applied and blended into each other with the brushstrokes.

Another possibility is to paint over a completely dry ground, allowing opaque or transparent colors to be added without altering the underlying layer of paint.

Oil is the ideal medium for obtaining a wide range of textured surfaces, such as these apples.

A texture created over a dry ground with a piece of cloth stained with ocher-colored oil paint.

Potential for Matter Painting

Oil not only allows the artist to delicately blend and superimpose colors in order to obtain visual textures but also to work with thick layers of paint in the style of matter painting, in other words, a stable and malleable modeling of forms on the canvas. The texture of oil paint can be emphasized using additives, meaning any kind of sand or ground mineral that does not oxidize: ordinary, washed sand; marble dust; oligist . . .

An example of the matter potential of oil; in this case no additives have been used.

Oil painting can represent the texture of skin using darker tones that gradually merge into lighter ones.

Value Painting and Gradations

Another of the many possibilities of oil paint for creating different textures that can be developed in the painting is modeling the forms by gradually merging the areas in shadow with the illuminated areas. This process, by which a dark color gradually merges into a luminous tone is called value painting, and the process by which a color is gradually transformed into another color is called gradating. If you study the small example here you will notice that the darkest color is used for the shape of the torso; then a lighter, more vivid color is added to the most luminous areas, blending each one into a gentle tonal gradation.

Superimposing and Blending Colors

The plastic qualities of oil paint are numerous.

Oil's pasty consistency and slow drying allows the artist to superimpose colors even while the paint is still wet. Its density

Oil can be drawn out, worked as impasto, superimposed on other layers of paint or blended with them to create any kind of texture.

also means it can be gently blended to form subtle chromatic gradations.

For creating the texture of highlights in oil painting, each material is treated in a different way.

A highlight on glass can be imitated with a simple white brushstroke, whereas a highlight on cloth may be painted using a blend of colors.

The Best Medium for Imitating Reality

Oil is the best medium that exists for enabling the artist to capture reality through the representation of texture. The dense consistency of this medium allows all types of plastic alterations to happen on the surface of the canvas. Notice how the forms and textures have been modeled in this painting by Audrey Flack.

Audrey Flack (1931). Family Portrait (1969–1970), oil on wood. Rose Art Museum, Waltham (Massachusetts).

MORE INFORMATION

• Texture and matter painting **p. 8**

PICTORIAL MEDIA: ACRYLIC

Although acrylic was used for industrial purposes during the middle of the twentieth century, it was not until the appearance of pop-art, Abstract Expressionism, and hyper-realist painting that it became a common item in artists' studios. Today it is, without a doubt, one of the most multi-purpose media that exist, given the advantages it offers over other media: it is soluble in water when wet, waterproof when dry, and is stable in terms of color and consistency.

Acrylic paint is ideal for quick brushwork; its fast drying time means colors can be applied quickly to create all kinds of textures.

Acrylic paint can be used thick or highly diluted.

Quick Drying and Instant Texture

Unlike oil paint, acrylic dries in a very short period of time, allowing the artist to superimpose colors or make corrections; this means an acrylic painting can be finished in a single session. Also, acrylic paint can be used in the same way as oil paint to form dense, pasty layers, or it can be thinned down.

Acrylic paintings can be developed very quickly, given the medium's fast drying time.

The Best Alternative to Other Pictorial Media

Acrylic can be used in a similar way to watercolor (for creating transparent effects and glazes) or to oil (when used in its dense form). Acrylic allows for painting that seeks to represent textures in a realistic fashion to develop very quickly. Whereas oil paint requires a long wait to create textures, sometimes even months for

the different layers of paint to dry, acrylic can be used to carry out works of high quality, similar to oil paintings, in a matter of minutes. In addition, the drying time for acrylics can be shortened even further by simply using a hair-dryer.

Finishes and Textures

The finishes you can obtain with acrylic are very similar to those available with oil paint, although it does tend to acquire a characteristically shiny effect when applied thickly. The textures obtained with acrylic, used in any pictorial style, can effortlessly imitate any surface present in nature.

Textures painted in acrylic can be either rough or completely smooth, depending on the additives contained in the medium. These additives are different kinds of ground minerals such as marble, alabaster, carborundum; dense, compact paint can also be obtained using a chemical thickener (catalyst).

Detail of a dense area of paint that allows the color of the canvas to show through.

Oil paint has been used to produce a perfect finish over an acrylic ground.

All kinds of opaque textures can be applied over a transparent layer of acrylic.

Ground for Other Pictorial Media

When a lot of preparatory work is necessary to create textures, acrylic is the perfect base for other slow-drying pictorial media such as oil paint. Being able to use acrylic as the ground for a given texture means that the painting can be well developed before the oil paint is applied. In other words, the chromatic range of acrylic combines perfectly with that of oil.

Acrylic Paint and the Airbrush

It is possible to represent any kind of surface using acrylic paint, however complicated it may seem; the most important factor is a sound knowledge of the technique and knowing how to apply this knowledge at the right moment. The metallic highlights on this

Chevrolet were painted with an airbrush; the resulting realism of the texture of the chassis is a close match to the sharpest of photographs.

Miquel Ferrón. Chevrolet, acrylic and airbrush.

PICTORIAL MEDIA: PASTEL

Although pastel would appear to fall into the category of drawing media,
it is undoubtedly the most direct and spontaneous pictorial media that exists.
When the artist makes use of all its technical potential, he discovers that behind its simple
appearance there is a wealth of different possibilities that can be as complex to use as oil or
acrylic. The textures that can be obtained with pastel can be as varied as those
obtainable with any other pictorial medium.

Creating Texture with the Fingers

Despite being a pictorial medium, pastel is generally applied dry, that is, it is not mixed on the palette and then applied to the painting. It is generally applied either in its stick form or with the fingers. The textures obtained with pastel are produced by alternating the use of the stick with blending with the fingers. The different textures that pastel can create are obtained by blending and gradating tones and colors or by directly superimposing masses of color on a surface that has already been colored or gradated.

Color and Form

Pastel, more than any other medium, is applied over an initial drawing. While oil can be used to construct the form of an object as the painting progresses, pastel, being a dry medium, needs the

A brief summary of the pastel technique.

The best way of blending pastel colors and modeling textures is using the fingers.

different objects in the composition to be present from the preliminary blocking in.

The ground color of a work done with pastel is usually related to the color of the paper itself.

The applications of pastel separate the areas of light and shadow from the color blend itself. This tonal difference allows the forms of the painting to be modeled and enriched using other colors.

Texture in Drawing and Painting

Textures created with pastel are modeled with the fingers or the blending stump and are retouched using impastos as if they were a drawing. In this way, the painting can be carried out as if it were a drawing while maintaining the characteristics of the medium itself.

Thus, textures painted in pastel can be divided into blended textures and graphic textures, a combination that can produce interesting effects which, depending on the artist's technique, can be highly realistic.

Edgar Degas (1834–1917), Pause in a Ballet Class. *Museum of Fine Arts, Denver. The textures of the tutus were created with an initial blending and then drawn in impasto.*

Competing with Reality

During the seventeenth century, pastel began to rival oil, especially in France. There were many artists who chose this technique for creating beautiful works of art that, as far as texture is concerned, were comparable to the most elaborate of oil paintings. Liotard was a master of this medium; his works were those that most rivalled the "queen of media" and, in many cases, the results he obtained far surpassed those of other artists famous for their mastery of oil painting.

Jean-Étienne Liotard (1702–1789), Supposed Portrait of the Countess of Coventry. Museum of Art and History, Geneva.

In this portrait, the painting was carried out as the drawing developed.

The Potential of Pastel

The most subtle of textures created with pastel are those that represent large, gradated masses of color, such as skies and seas; however, it can also produce fine results when re-creating the texture of skin or hair, subjects that require highly detailed work.

As with all pictorial media, pastel requires that the artist have solid technical knowledge so as to obtain good results, especially when representing relatively complex textures.

Virtually any subject is suitable for pastel painting.

MORE INFORMATION

- Mixing media to obtain textures **p. 36**
- Cloudy skies with pastel **p. 72**

PICTORIAL MEDIA: WATERCOLOR

Without a doubt, watercolor is one of the most complex pictorial media that exist.
This is partly because it uses water as its medium, an element that lacks consistency and
acts as the vehicle for the pigment and the gum arabic. Watercolor is not as intuitive
a medium as oil or pastel because it lacks the color white. The brighter tones must be
obtained from thin transparencies, while on the other hand, its very transparency means
that bright, light colors cannot be superimposed.

The Texture of Water

The water present in watercolor paint has a particular effect on the paper. There are basically two ways of painting with watercolor, which can, of course, be combined to achieve other effects. One is to wet the paper first with water or with a watercolor wash; when the paint is applied it has a strong tendency to spread over the wet area, an effect that can be used to create gradations or suggest textures such as skies, smoke, or surfaces that will later be retouched. The other method is to paint over a dry ground, which leads to precise brushwork that can be used to faithfully represent all kinds of forms.

Watercolor painted on dry paper.

Watercolor painted on wet paper.

The two previous techniques combined; the more precise forms and masses were painted on a dry ground.

Paint and Light

The texture of any object painted with watercolor depends on both the light of the object itself, and that of the paint on the paper.

The artist must study the texture of the object before starting to paint on the paper so as to analyze the highlights and other features of each surface.

Light in watercolor painting disappears as successive glazes or layers of paint are superimposed on the dry surface. Starting with highly transparent colors, additional layers create different tonalities as well as the form and texture of the objects in the composition.

Different glazes create the texture, volume, and highlights.

Textures: From Light to Opaque

When working with watercolor, textures can be obtained in many different ways depending on the material available or the atmosphere you want to achieve.

All watercolor brushmarks create a texture on the paper, so by combining the dry and wet watercolor techniques and by using different materials previously soaked with watercolor paint, you can obtain all kinds of textures. These can be applied to the painting so that the lightest colors are those that form the background.

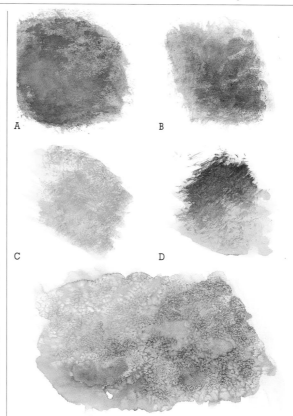

Watercolor textures obtained using a sponge (A), a cloth (B), blotting paper (C), and an almost dry brush (D).

MORE INFORMATION:
- Additives for creating textures **p. 32**
- The paper, its grain, and the results of different processes **p. 40**

A texture obtained by sprinkling salt on the wet, watercolor ground.

Realistic Textures

Watercolor is an excellent medium for obtaining all kinds of effects, though it does require a sound knowledge of the techniques involved in order to create perfect textures; this explains why many Naturalists use watercolor to paint features and animal skin. It is very important that every learner experiment with creating different textures on paper, or simply dabble with the different possibilities, for they will come in handy when actually painting from real subjects. Notice that the textures we have created in this chaper could be used to paint rocks, clouds, skin ...

Learning from the Masters

One of the best ways of learning and practicing with watercolors to create different effects is to observe how the great masters used them in their works. Every detail of a masterpiece reveals numerous techniques that a learner can use to his own advantage when painting. They can be copied or studied in museums or in reproductions from books.

PICTORIAL MEDIA: CRAYONS

Although it is one of the least known pictorial media that exist, crayons allow the artist to create a vast range of different effects, not to mention the possibility of combining them with other similar and dissimilar media. Crayons can be mixed with oil or pastel; however, they repel water and are therefore unique for creating textures by means of masking certain areas of the paper. Since they are a repellent for both watercolor and inks, crayons protect the surface of the paper by making it impermeable.

A

The Finest of Textures

Crayons provide the artist with a wide variety of potential effects due to their plasticity.

This means that crayons can be transformed either before or after being applied.

If you paint with the crayon directly on the paper, you can obtain a fine texture or one that completely seals the grain of the paper, canvas, or wood.

B

It can also be melted while it's being applied, for its low melting point allows the artist to mix or blend colors by simply drawing his fingers quickly over the painted surface.

Several layers of crayon can be superimposed without having the colors mix together.

C

Usual Methods

(A) Crayons can be used like pencils to mask areas of a surface and thus prevent ink or watercolor from penetrating it.

(B) They can also be used in a similar way to a transparent wax stick, creating interesting colors and textures if ink is later applied to the surface.

(C) Crayons may be dissolved with turpentine.

(D) They can also be melted with the heat of a flame.

D

Different textural effects obtained using crayon.

Crayon mask painted over with ink.

Melted and Dissolved Crayon

Crayon melts easily with the warmth from the fingers if it is applied by rubbing one color over an underlying tone. This technique is useful for creating the texture of highlights on fruits or metal. If fast melting is required, a direct source of heat can be used, such as a hair-dryer or a flame. This process will cause the crayon to liquefy almost immediately before being shaped using a brush or palette knife. When a much more liquid or thin quality is needed, the paint can be dissolved using a brush dipped in turpentine.

The texture of the fruit was obtained by rubbing the white with the fingers until it melted on top of the red.

Masking and Crayons

Masking is one of the best ways of creating different textures when used with mixed media. Masking to obtain the effect of a "negative" involves understanding the highlights and points of light in advance and leaving the darker colors for a later stage in the development of the painting. If ink or watercolor is used, these darker colors will penetrate the areas not masked with the crayon.

Compatibilities, Incompatibilities, and How to Make Use of Them

Crayon is the only pictorial medium that allows its incompatibilities to be employed with other media. Many of the different possible effects that can be achieved with crayon are obtained by scratching the surface, sgraffito style, or by using the crayon itself to protect areas against water-based media, such as ink or watercolor.

Remember that a surface painted with crayon cannot be painted over with tempera, watercolor, or ink; however, it can be painted with oil or pastel.

MORE INFORMATION
· Mixing media to obtain textures **p. 36**

Combination of techniques: sgraffito, white reserves, and working in negative.

ADDITIVES FOR CREATING TEXTURES

Additives are of great importance in matter painting, for they not only lend body
to the pictorial medium itself, but also add a new dimension to the medium that can be used
to express all kinds of matter textures. Each pictorial process allows different kinds
of materials to be used, although there are additives that can be used in all
painting processes (except those that are too liquid such as watercolor)
or with dry media such as pastel.

Matter Texture

Interestingly enough, additives bring a foreign component to the pictorial medium that causes it to acquire a texture that is quite different to the kind it would otherwise have when dry.

There are two kinds of additives: ground minerals and chemical products; both alter the way the medium becomes dense, the same way thickeners or gesso behave with acrylic.

An example of matter painting; this oil contains no additive whatsoever; the texture has been created by using the palette knife.

Sand and Marble Dust

Washed sand is suitable material for mixing with oil and acrylic because it thickens the paint, lending it an abrasive finish.

Marble dust is often ground very finely; depending on how fine it is, when mixed with oil, it will increase the medium's volume and make it more compact, giving it a sandy texture that does not lose its moldability.

There are many effects that can be obtained using marble dust, including even surfaces with a lot of volume and others in which the grains of sand or marble are covered with the paint but

Different kinds of washed sand; since they are not pigments, they can be mixed with oil or acrylic.

retain their textural effect to the touch.

The interesting results obtained by adding marble dust of a specific color also are to be noted.

Other Additives

These are substances that increase the volume of the medium

without creating a different texture to the touch or altering the color. The usual additive for oil and acrylic, known as Spanish white or Paris white (ground calcium carbonate, washed and refined) produces good results; lithophone (zinc sulfate and barium sulfate) is also used.

These additives are inert pigments that can be used to increase the volume of the paint,

Marble dust can be used to obtain a highly compact, stable texture for oil and acrylic painting.

Pictorial Media: Crayons
Additives for Creating Textures
Mixing Media to Obtain Textures. Acrylic and Oil

33

while at the same time making whites more opaque.

Chemical additives also exist for oil that can thicken it considerably and increase its density, such as painting paste.

Thickeners and Acrylic Gels

In addition to additives, the texture of acrylic can be altered by using acrylic additives that thicken it.

Some of these thickeners come in the form of dense, colorless gels that serve to increase the density of the medium. Others act as a catalyst on the resinous medium, causing it to thicken.

Another way to lend body to paint is to prepare the pictorial ground with acrylic pastes that can be altered to produce a textured surface before you actually start to paint.

Painting paste is mixed with oil paint to increase its density.

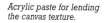

Acrylic paste for lending the canvas texture.

MORE INFORMATION

- Texture and matter painting **p. 8**
- Matter painting **p. 14**

A Master of Matter Painting

Bengt Lindstrom is a contemporary artist who has taken great advantage of texture painting, mainly with oil. His paintings have layers that measure up to several centimeters thick, and are manipulated and modeled using a wide range of tools.

Bengt Lindstrom (1925), Oil on Panel. Private Collection.

MIXING MEDIA TO OBTAIN TEXTURES. ACRYLIC AND OIL

Mixed media refers to a pictorial method that combines at least two different painting media in a single piece. The compatibility of pictorial media is always determined by an unchanging norm in painting that states that "thick should be painted over thin," in other words, oil is considered thick and acrylic thin. Furthermore, oil paint is considered thinner the more turpentine it contains, as opposed to pure oil paint. When painting textures, acrylic paint dries quickly while oil produces a high quality finish.

The subject has been drawn and then the acrylic colors added without varying the tone.

Acrylic paint dries very quickly and, if no additives are used, it will create a ground that is smooth. The ground will be flat as far as color is concerned if no other colors are painted over it.

Starting a painting with a completely flat surface makes it easier to correct details when using oil.

If, however, acrylic has been used to texture the surface, the brush marks in oil will be affected by this feature of the painting.

This preliminary work in acrylic should create a flat painting, as far as both color and texture are concerned.

A Quick Start

In order to superimpose two pictorial media such as oil and acrylic, the artist must first anticipate the result of the first layers of acrylic before beginning to paint in oil.

Acrylic is a medium that enables the artist to apply large masses of color very quickly so, if the sketch of the subject has already been drawn, the resulting textures will be obatined sooner than if the artist attempts to find the color through the form.

Flat Areas of Color

Creating flat areas of color will allow the artist to build a foundation on which oil will be employed to obtain all kinds of textures.

Thanks to the flat ground, the oil can be worked extensively in order to create the desired textures.

MORE INFORMATION

- Texture and matter painting **p. 8**
- Pictorial media: Oil **p. 22**
- Pictorial media: Acrylic **p. 24**
- Additives for creating textures **p. 32**

Additives for Creating Textures

Mixing Media to Obtain Textures. Acrylic and Oil

35

Mixing Pastel and Other Media

Applying Oil Paint

When applying oil paint over an acrylic ground, it is important to follow a series of simple, unchanging rules that can be applied to any kind of mixed media painting that includes oil.

Once the acrylic ground is completely dry, the artist can begin to apply the necessary layers of oil paint to create the texture of the painted forms.

The preliminary layers of oil should be thinner at first, which can be done by simply dampening the brush in a little turpentine and then mixing the colors on the palette.

When painting over fresh oil, slow drying colors such as black should be avoided in the underlying layers, for this could cause the outer layers of paint to crack when dry.

Subjects

The use of mixed media that require a quick drying time in the initial phases of the painting, as it is the case when oil is applied over acrylic, is particularly suited to subjects that are considered appropriate for quick paintings, such as travel sketches, quick sketches of a model, or quick still life painting.

Since these two media can quickly be superimposed, it becomes simpler to paint the different textures, a process that otherwise could be a painstaking task.

Although they may have no need to use the combined acrylic-oil technique for quick painting, there are many artists who use it to paint all kinds of subjects, using each medium to its best advantage.

The colored ground is almost complete, using acrylic colors.

When the acrylic is dry, oil is applied and either blended with the ground colors or to form transparencies over the underlying layers.

Quick Painting and Mixed Media

What is generally called quick painting has many enthusiasts, both professional and amateur painters included. If you practice this kind of painting, even if only as an exercise in interpretation, you can discover many pictorial resources that can be used in all kinds of work. For example, knowing how to combine different techniques can enable you to create textures with acrylic that are impossible with oil, such as thick layers of transparent paint.

Acrylic can be used to obtain textures that are impossible with oil, such as thick layers of transparent paint.

MIXING MEDIA TO OBTAIN TEXTURES. PASTEL AND OTHER MEDIA

There are many ways of working with mixed media. Pastel offers many different uses when combined with other media; it can be mixed with virtually any other medium because, being dry, it can be absorbed by other binding compounds. Oil pastels are sold that enable the artist to work with a brush or with oil as the solvent dissolves the pastel that has already been applied to the surface.

Combining Techniques

Dry media (pastel, charcoal, sanguine, crayon) are among those that are most suitable for combining, for they can be blended together in different ways to create different effects, without them repelling each other.

Dry media can also be combined with wet ones such as acrylic, oil, watercolor, or gouache, depending on the proportions used.

In any case, the artist should follow the unmovable rule of painting which is to paint thick over thin.

Pastel is particularly well suited to being combined with all other types of media, even with oil paint (an oil-based medium) or gouache (a water-based medium).

Any pictorial medium can be applied over this pastel color.

Wet and Dry Techniques

Dry techniques can easily be combined; however, when they are combined with wet techniques, the solvent contained in the latter will either be accepted or repelled by the former.

If gouache, oil, or acrylic is applied over pastel, they will blend together; however, if crayon is the first medium used, only oil or acrylic can be applied over it. (The crayon may repel the acrylic until the water has totally evaporated.)

Different pictorial media can be combined to produce textures, although the rule of thick over thin must always be followed. One of the easiest media to combine is pastel.

Investigating Textures

For an artist who is investigating how to develop textures, it is enlightening to consider how children paint; they take the paint with their hands, and use all kinds of painting instruments to rub, smudge, or simply stain the surface.

A child with a can of paint is often more adventurous and innovative than an adult, for he has not been conditioned in a certain way, but instead simply paints for his own amusement.

Mixing Media to Obtain Textures. Acrylic and Oil
Mixing Pastel and Other Media
Textures: Pastel, Crayon, Gouache, Watercolor

37

Naturally, if watercolor is applied over crayon, it will be repelled and deflected to the areas where there is no crayon, creating a negative image when dark watercolors are applied over light-colored crayon.

Combinations of Oil and Turpentine

Both pastel and crayon are media that combine well with oil and can be altered by the turpentine that oil paint contains as a solvent.

Blending media is an excellent means of obtaining all kinds of textures, given that the pastel marks can be combined with oil impastos.

When mixing pastel and oil it is best, though not essential, to use oil pastels; these are pastels that have already been mixed with a certain amount of oil making them much easier to combine with other media.

Turpentine allows pastel to be diluted in oil in such a way that it can be spread over the surface of the paper like a wash.

This flower was painted using oil pastels and finally retouched with oil paint.

MORE INFORMATION

- Pictorial media: Crayons **p. 30**
- Additives for creating textures **p. 32**
- Textures obtained with pastel, crayon, gouache, and watercolor **p. 38**

Different textures obtained by blending crayon and pastel.

Improvisation and Practice

Combining pictorial media provides the artist with many different forms of expression that cannot be developed using just a single medium. The painter's studio should be a place to experiment and devote time to investigating how different media can be used together. This investigation will often produce surprising results that the artist should make a note of and remember so as to use them later on and make them part of his technical background.

Two examples of how to apply oil pastels; blended or dissolved in turpentine.

TEXTURES OBTAINED WITH PASTEL, CRAYON, GOUACHE, AND WATERCOLOR

Although different pictorial media such as crayons, gouache, and watercolor appear to be incompatible, they actually combine particularly well. Dry media such as pastel and crayon can be easily mixed to obtain a wide variety of effects. Crayon dissolves easily in turpentine, enabling the artist to create gently gradated textures; when mixed with other water-based media, crayon tends to repel them, creating a plastic effect.

Colored ground painted in pastel.

Dissolving Crayon with Turpentine

The most interesting effects that can be obtained with crayon result from dipping a brush in turpentine and passing it over a painted surface. The effect is very similar to watercolor; however, the difference is that with crayon, completely opaque colors can be used in the parts of the painting

After blending the color, the highlights of the skin and the fabric of the blouse have been painted using white crayon.

Using Pastel and Crayon Together. Skin and Fabric

Pastel can be applied as a ground for crayon or one to be painted over. Easy to blend and correct, pastel produces textures that are simple to alter, for you can apply thick impastos or blend the medium to merge it with the color of the paper. Textures obtained using pastel can imitate the velvety quality of skin and fabric, by blending tonalities to which other tones and highlights can be added.

Crayon is more stable than pastel and can be used to alter the highlights of the skin and fabric by taking advantage of the contrast produced by a white color over a different colored paper.

where solvent has not been applied. Another possibility crayon offers is being able to apply light colors over dark ones, something that is not possible with watercolor. The textures obtained by dissolving crayon with turpentine makes it easier to paint highly transparent and luminous surfaces, over which thick layers of paint can then be applied.

Negative Images with Crayon and Watercolor

Another effect that can be obtained using crayons results from applying watercolor to those areas not already painted in crayon. Next, using the stub of a candle, the highlights of flowers and other objects can be painted and finally, if a dark color is applied over the painting, the waxed area will stand out. Other luminous colors can be used in the same way, for the watercolor cannot penetrate these areas that are covered in wax.

The Right Texture

By combining techniques that provide interesting results, such

The effect of the grass, water, and mist in the background has been achieved by dissolving crayon with turpentine.

as the ones described above that are based on the incompatibility of crayon and watercolor, a great variety of textures and highlights can be obtained. These techniques, however, should not necessarily be used over the entire painting; indeed, moderate use of these different techniques make the results far more interesting. In any case, the resulting effect is not always the same for it depends on the grain of the paper, on how much wax crayon has been applied, and on the final brushwork with the watercolor.

The luminosity of crayon and watercolor is a constant source of surprise.

Different effects from negative images.

The Texture of Encaustic Painting

Encaustic painting, which dates back to ancient Greece, is based on the use of hot wax.

Encaustic paint is obtained by mixing dry pigments, white beeswax, and Dammar varnish on a hot palette, and then applied while hot using a brush or palette knife.

Josep M. Ávila, Abstract Composition, encaustic on wood.

MORE INFORMATION

- Pictorial media: Acrylic **p. 24**
- Pictorial media: Crayons **p. 30**
- Mixing media to obtain textures. Pastel and other media **p. 36**

THE PAPER, ITS GRAIN, AND DIFFERENT PROCESSES

The surface is especially decisive for obtaining different textures. Paper is one of the most commonly used surfaces since, with the exception of oil, all media are perfectly apt for painting on it.

Of the many types of paper available, we will only mention a few varieties and their basic characteristics. An artist should always take into account the different characteristics of the paper when trying to achieve specific textures.

The type of paper to use depends on the process one intends to follow.

Fine Grain Paper

The texture of paper depends on how it is manufactured. Its surface is smoother if, during the manufacturing process, the pulp (shredded cotton soaked in water) is collected in a fine wire mesh and then pressed suitably. Fine grain paper is suitable for drawing on since the pencil encounters no obstacles as it moves across the surface. All paper undergoes a process of sizing and bleaching; when the sizing has acquired enough strength, the paper can be used for watercolor painting regardless of how fine it is. Excessive sizing, however, prevents the paper from absorbing water, and therefore it tends to be thick and is more suitable for acrylic paints. The grain of paper is determined by its weight rather than its thickness.

Fine grain paper.

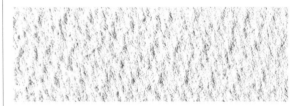

Medium grain paper.

Medium Grain Paper

Medium grain paper is probably the most commonly used, since any medium can be applied to it. Different pictorial media adapt easily to the texture of the paper. Medium grain paper facilitates a more detailed work similar to what can be achieved on fine grain paper. The texture of this type of paper can also be incorporated into the work itself. When the artist is painting with dense paint like acrylic, it is essential to work on nothing less than 250g paper, otherwise the paper will crease due to the dilation of the pores and the ensuing contraction of the paint. If you

paint with wet media, it is advisable to dampen the paper beforehand with a sponge and have it firmly attached to a rigid support.

Two texture tests; the first, on fine grain paper; the second on rough grain paper.

Rough Grain Paper

Rough grain paper is the type that is most obvious to the eye for its surface is heavily textured.

Rough grain paper is only recommended for certain types of work in which the texture of the paper itself forms an important part of the painting.

Of the many rough grain papers that exist, the most attractive for painting on are the handmade types.

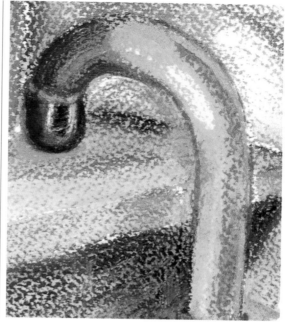

Rough grain paper allows the artist to incorporate its features into the painting.

MORE INFORMATION

- Drawing textures **p. 60**

Example of rough grain paper.

Handmade Paper

Handmade paper is without a doubt the most attractive type on the market. As its name indicates, this paper has been manufactured entirely by hand, a fact that characterizes its finish, featheredge, and texture. This kind of paper is far more expensive than the rest; nonetheless, thanks to its rich and attractive texture, it is particularly apt for certain types of work.

The Texture of the Surface and Its Importance in the Result

It is essential that the artist choose the right type of paper for his work, since its texture will influence the final outcome of the piece.

Compare the difference between these two results.

Different types of handmade paper: the most attractive of all surfaces.

MATERIALS FOR TEXTURES IN WATER-COLOR: SALT, TURPENTINE, BLEACH

Although every process allows one to create a series of different textures based on the manipulation of the medium, there is a diverse range of textures that do not depend solely on the way the brush is handled but rather on the use of additives. Different materials can produce the dissolution or concentration of pigment in specific zones of the painting. Watercolor is one medium that is particularly apt for achieving such alterations with the aid of additives.

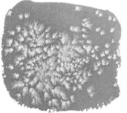
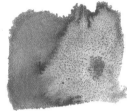

Two different uses of salt on watercolor paint.

MORE INFORMATION

- Texture and matter painting **p. 8**
- Pictorial media: Watercolor **p. 28**
- Textures obtained with pastel, crayon, gouache, and watercolor **p. 38**

Areas of the paper can be kept free of watercolor by applying turpentine to them.

Using Salt

Salt is one of the best absorbents of dampness; for this reason, if it is applied to a surface wet with watercolor, the grains will tend to absorb the liquid and dissolve as a result. This effect can cause the grains to clean the surface on which they have been placed or to absorb the water until the watercolor pigment is concentrated in the area where the absorption has taken place.

Salt is often used to create the texture of rocky terrain and for painting buildings whose facades have a pronounced, grainy texture.

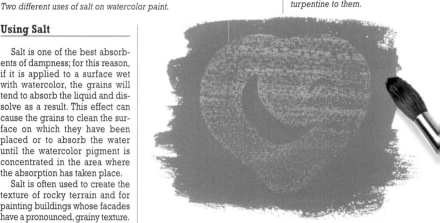

The texture of this wall has been created using salt and retouched afterward.

Turpentine

Turpentine can be used to obtain textures with watercolor by sealing the pores of the paper with a very light, greasy film, making the surface water-resistant without damaging it. When turpentine is applied directly on white paper, it is possible to protect forms and objects from being covered in paint. Furthermore, turpentine can be applied over a surface recently painted with watercolor with the purpose of causing the watercolor to open up and create highly original and effective textures for painting treetops, embankments, or earthen walls.

The Paper, Its Grain, and Different Processes
Materials for Obtaining Textures in Watercolor
Modeling Paste and Pastel

43

An application of turpentine over freshly painted watercolor.

A Resource for Every Occasion

There is a specific technique appropriate for each occasion; in other words, it is senseless to use a technique when there is no need for it. With each textural effect in its rightful place, the work will acquire an interesting plastic value.

The Use of Water for Blending Areas and Creating Backgrounds

When preparing backgrounds with watercolor, water can be used to obtain interesting effects by blending the colors applied to the paper.

A textural effect allows us to execute unusual backgrounds by merely washing the paper once it has dried or by dampening it with a sponge.

Highlights with Bleach

Since there is no white in watercolor, highlights of textures are obtained by using the white of the paper itself.

Despite the fact that watercolor is irreversible once it has been applied, it is possible to open up white areas and correct textures; this however, depends to a great extent on the type of paper used. If the paper is of high quality, all the artist need do is make several sweeps of a brush loaded with a solution of 50% bleach and 50% water over the area to be cleaned.

The effects of washing over watercolor with water once it has dried.

Dry watercolor can also be washed with a sponge.

This effect was achieved with water and bleach. It is advisable not to carry out this procedure with a good brush.

MIXED MEDIA FOR OBTAINING TEXTURES. MODELING PASTE AND PASTEL

Modeling paste allows the artist to give texture to forms that can later be painted any color. This kind of paste is not difficult to handle and is extremely attractive both for preparing the canvas and for defining specific forms on it. Furthermore, since it is transparent you can obtain body without dirtying the color. The acrylic origin of modeling paste facilitates mixing the medium with pastel.

Uses and Application of the Material

Modeling paste can be applied both with a brush and with any other drawing or painting utensil. All kinds of textures, both modeled or scratched, can be created over it. Furthermore, due to its acrylic origin, this material dries quickly, thus allowing a piece to be completed in a single session. Nonetheless, it is important to bear in mind that, with the exception of acrylic paint, the thicker the material you use is, the longer it takes to dry.

Acrylic modeling paste, pastels, acrylic paint, and other accessories for painting.

We begin to cover the forms as if we were modeling them.

The forms are modeled with a palette knife.

Starting a Texture

Modeling paste can be applied to any non-greasy surface; that is, on paper, canvas, or wood that has not been primed or painted on beforehand. Since modeling paste is white when applied, the artist can momentarily forget about color and concentrate on obtaining textures and modeling forms. In this example, we are modeling the shape of a tennis shoe that has first been drawn. Each one of the forms is treated as independent, and each part occupies a different plane in the painting. The modeling paste is applied with the brush, gradually following the direction of each and every one of the forms in the drawing until they are entirely covered.

Modeling Texture and Color Applications

Once the form that one wishes to give texture to has been covered, the real modeling work begins. The palette knife is an indispensable tool for modeling, since it can used to smooth or flatten areas that may require it; other areas, such as the laces, can be brought out by building up a thin strip of paste to create relief. Once the figure's modeling process is complete, we can begin to create textures with all kinds of tools: a cutter, a graphite pencil, the tip of a brush ... Once the paste has dried, we can draw over it with a lead pencil and apply the pastel colors.

Once the modeling paste is dry, the surface can be drawn on with pencil and painted with pastel.

Touching-up and the Finish

Pastel colors are ideal for covering a textured area regardless of how rough it may be. The pastel is applied in stick-form and is easy to spread with the fingers.

Effects as Defects

One of the most common mistakes made is to abuse the use of effects in a painting. These should be used to enhance the work but not become the protagonist. Excess effects make the painting boring: the effects must be used as a tool or technique, but never as the final and main focus.

MORE INFORMATION

- Pictorial media: Acrylic **p. 24**
- Pictorial media: Pastel **p. 26**
- Mixing media to obtain textures. Pastel and other media **p. 36**

Since the shapes of the shoe are clearly structured and dry, all of its areas and planes can be made out easily. The dark colors are blended with white by rubbing over them with the fingers.

Once the main highlights are in place, we can stain the background with a very transparent ocher wash, but since the texture must reflect the surrounding colors, touches of the background color are added to the shoe.

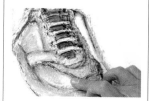

The color is easily blended using the fingers.

Ocher is applied to reflect the surrounding color and integrate the color of the shoe.

TEXTURES WITH AIRBRUSH ON SMOOTH GEOMETRIC BODIES (1)

The airbrush can be used to achieve all manner of textures, including almost photographic results when applying highlights to smooth bodies in the work. The airbrush is not difficult to use; however, it does require plenty of experience from the artist in order to create interesting works.

By studying different forms, the way light responds on different planes, and with the aid of geometric elements that will make it easy to paint any shape on paper, it is possible to develop a technique for resolving the characteristically smooth textures of geometric bodies.

The Use of the Airbrush, Materials, and Components

The airbrush is a tool that sprays paint, contained in a tank, with precision.

Airbrush painting is not complicated but requires practice and a certain amount of expensive and sophisticated equipment.

The airbrush unit requires a good quality compressor and a connecting rubber tube. Each one of these components is shown in the accompanying illustration. It is important for the novice to become familiar with the different parts of the equipment before using it.

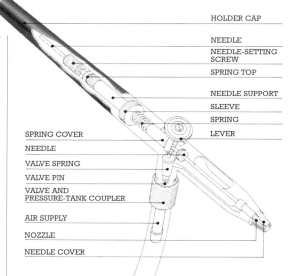

HOLDER CAP
NEEDLE
NEEDLE-SETTING SCREW
SPRING TOP
NEEDLE SUPPORT
SLEEVE
SPRING
LEVER
SPRING COVER
NEEDLE
VALVE SPRING
VALVE PIN
VALVE AND PRESSURE-TANK COUPLER
AIR SUPPLY
NOZZLE
NEEDLE COVER

The different parts of the airbrush.

MORE INFORMATION

- Imitating different types of surface **p. 12**
- Textures with airbrush on smooth geometric bodies (2) **p. 48**

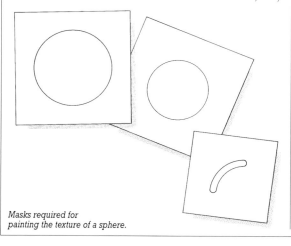

Masks required for painting the texture of a sphere.

Reserving, Applications

In order to paint textures with the airbrush, it is essential to know how to use masks to reserve zones. This is necessary to prevent the spray from covering or invading areas that you wish to keep white or paint another color. Masks can be made from paper or with adhesive tape, the latter being the most precise since once it is attached to the paper, there is no danger of painting the reserved area. A mask cut out of paper or cardboard, on the other hand, is more mobile and can be held at a distance from the paper in order to vary the definition of the contours.

Airbrush process of a sphere using the previous masks.

The Sphere

Three masks are required to paint a completely smooth sphere. The first is used to paint the general mass of the shape. The second is for the smaller circle, and facilitates the relatively precise concentration of paint in the sphere's central zone. The paint should be applied from a distance to obtain a subtle gradation. The contrasts between the areas of light and shadow are intensified before the circular masks are removed. Last, the third mask, which allows us to reserve the highlights, is placed in the space between the first and second masks. This area is airbrushed with white until the necessary highlights are achieved.

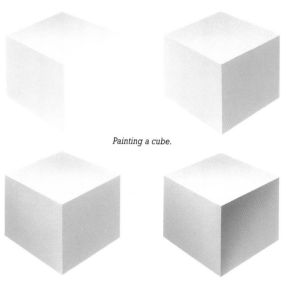

Painting a cube.

The Cube, Different Planes

The cube is the easiest geometric shape to paint because the masks employed can be identical. The only important question is knowing how to distinguish the intensities of the tones applied according to the light reaching each one of the cube's faces. Each plane receives a different degree of illumination. It is important to remember not to hold the airbrush too close to the paper in order to avoid highly linear applications.

Good Example of Airbrush Painting

Many illustrators use the airbrush in their work to produce extraordinary textural results. Note the combination of smooth textures with others that create the illusion of earth.

Illustration painted by Miquel Ferrón, a fine example of creating different textures with the airbrush.

TECHNIQUES AND PRACTICE

TEXTURES WITH AIRBRUSH
IN SMOOTH GEOMETRIC BODIES (2)

In order to create the different textures of metallic surfaces, it is essential to understand, and be able to represent on paper, how the surfaces of simple geometric forms behave when exposed to light. The airbrush allows the artist to paint uniform shading and subtle tonal variations, both with the use of masks and freehand.
Studies of the model's volume carried out with the airbrush permit a greater understanding of this subject and can be applied to any other pictorial medium.

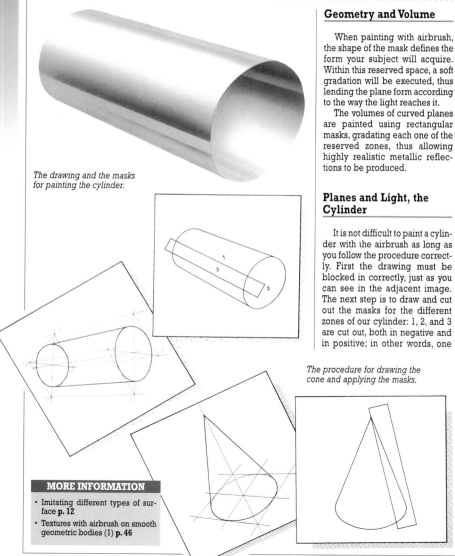

The drawing and the masks for painting the cylinder.

The procedure for drawing the cone and applying the masks.

Geometry and Volume

When painting with airbrush, the shape of the mask defines the form your subject will acquire. Within this reserved space, a soft gradation will be executed, thus lending the plane form according to the way the light reaches it.

The volumes of curved planes are painted using rectangular masks, gradating each one of the reserved zones, thus allowing highly realistic metallic reflections to be produced.

Planes and Light, the Cylinder

It is not difficult to paint a cylinder with the airbrush as long as you follow the procedure correctly. First the drawing must be blocked in correctly, just as you can see in the adjacent image. The next step is to draw and cut out the masks for the different zones of our cylinder: 1, 2, and 3 are cut out, both in negative and in positive; in other words, one

MORE INFORMATION
- Imitating different types of surface **p. 12**
- Textures with airbrush on smooth geometric bodies (1) **p. 46**

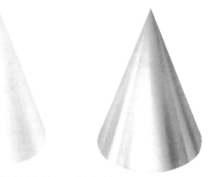

The procedure for painting the metallic texture of the cone.

mask is used to cover the form itself, while the other is used to enclose it. Zone 1 is painted without going into zone 2; then zone 1 is gradated and darkened using mask 3; finally, zone 2 is painted after having reserved zone 1 and its surroundings with the corresponding masks.

Study of a Cone

The cone is painted in a similar way to the cylinder; however, airbrushing and solving the shape of the cone is far simpler.

Only two masks are required, one for isolating the form of the cone and the other for reserving the points of contrast from the vertex to the base. In this case, the positive mask is not necessary, since we will only airbrush the interior of the form.

The entire contour is painted very lightly (ensuring that the center is not stained), and one of the sides is intensified in order to give it the

sense of volume. The highlights are added with the aid of a straight mask.

Composition and Geometry

Based on the simple forms that we have shown over the last few pages, it is possible to paint all manner of compositions.

By this stage, you should be aware of how easy it is to obtain the smooth forms of metallic surfaces with the airbrush, a technique that facilitates the creation of much more complex forms.

Study of the geometric forms of a still life.

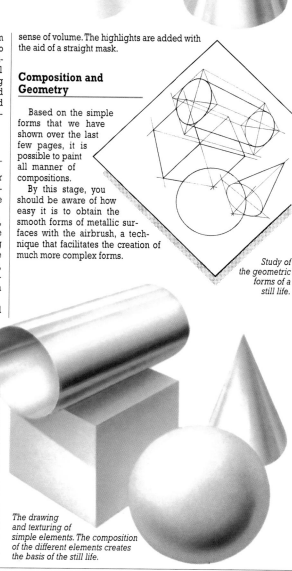

The drawing and texturing of simple elements. The composition of the different elements creates the basis of the still life.

HIGHLIGHTS IN HYPER-REALISM

Hyper-realism, also known as photorealism, is characterized by the almost
photographic way in which details are captured. In photographs of shiny objects,
especially metallic objects, each gleam or reflection can be easily made out.
If these zones are clearly marked out in a preliminary drawing,
they will be easy to paint.

Drawing Highlights

Before painting a hyper-realistic image with highlights and textures that contain sharp contrasts of light and shadow, we should always trace the image. This can be done either directly over the model or with the aid of a slide projector. When it comes to tracing or copying the model, special attention must be paid to the highlights on the object's surface. Shadows and tonal values should not be painted; however, the highlights should be outlined in order to obtain small enclosed islands that will later be protected when you begin to paint the surface. Remember to ensure that the drawing is clear and precise enough before you start applying color.

A Texture that Imitates Reality

Highlights not only exist on smooth surfaces like metal or glass, they are found on any surface subjected to the effects of light.

When you airbrush a specific surface, a particular type of texture is produced that results from spraying the paint with compressed air. This texture may be completely smooth or have a granulated appearance, depending on the distance from which it was sprayed and the amount of air pressure used. By using masks to reserve the zones to be left intact (meaning unpainted), we can superimpose several layers of paint to imitate textures such as those of ice-cream, bread, etc. The texture should first be tested on a piece of scrap paper to ensure it is correct before

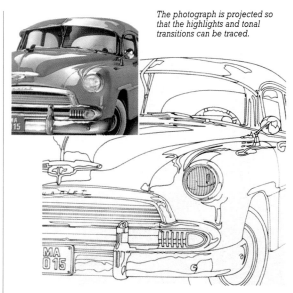

The photograph is projected so that the highlights and tonal transitions can be traced.

applying it to the definitive work. Also, the artist should always remember to study the direction of the light beforehand.

MORE INFORMATION

- Imitating different types of surface **p. 12**
- Textures with airbrush on smooth geometric bodies (1) **p. 46**
- Textures with airbrush on smooth geometric bodies (2) **p. 48**
- Highlights with oil **p. 68**

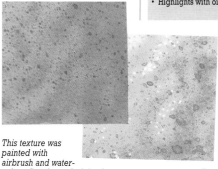

This texture was painted with airbrush and watercolors; first the underlying layer, then the darker mottled one.

Practical application of the previous texture.

Over this perfectly defined drawing, the main highlights have been neatly reserved with adhesive tape. The smaller glints are done with touches of masking fluid.

The Strongest Highlights on Glass, Varnish, and Different Metals

The brightest highlights are the result of sharp color changes in the contrast of those parts of the surface where the light bounces off.

When painting a hyper-realistic texture, it is important to start by drawing a sketch that simplifies each one of the different zones of light as much as possible. These bright areas, which correspond to the color white, are reserved with tape or with masking fluid, the latter being much easier to handle since it is applied as if it were paint. When the fluid is dry, the surface can be painted over without the danger of the paint

The result of a painting carried out with different reserving techniques.

penetrating the reserved section. Last, once the paint is dry, the masking fluid can be removed with an eraser or by carefully rubbing it with a finger. Once the fluid is removed, the zone will be completely white or whatever color was originally reserved.

Simplifying Textures

Even the most complex highlights can be painted easily if you define them beforehand.

There are many techniques for painting such textures, regardless of the medium you are working with. If you choose to paint with oil, acrylic or pastel, the different layers of colors can be

superimposed without any problems. If, on the other hand, you are using watercolor, these spaces will have to be left unpainted.

The textures must be concise and simplified; in other words, although there may exist a host of details within a specific mass of color that represents a highlight, these must be simplified as much as possible.

Controlling the Realism of Your Painting

The shiny nature of metal, as well as the effect of highlights on other materials, must be evaluated before they are included in the painting so as not to overwhelm the composition. If you heed this advice, you will obtain more spectacular results without falling into the trap of excessive effectism.

These stripes of color painted with acrylic can be used to represent metal.

THE TEXTURE OF GRASS

The texture of grassland depends to a great extent on the humidity and luxuriance of the subject. When you paint the model from a distance, the details become simpler, and the texture can be translated into one or various patches of color. Terrain painted from closer up, on the other hand, has a host of details and highlights that need to be captured through brushstrokes or patches of color.

Painting Tall Grass

When painting tall grass it is important to bear in mind how its luxuriance produces a very deep shadowy effect in its backgrounds.

Clumps of very high, compacted grass can be painted with a dark color and several violet notes. By starting with the darkest zones, we can leave the green grass to the last stage of the work.

When it comes to painting tall grass, the background of the painting is as important (if not more so), than the details.

Thus the background is painted with a couple of wide brushstrokes, adding more touches of violet to the foreground and reserving the other zones where, with the aid of a fine brush appropriate for outlining, green and yellowish tones will be applied to create the brightest blades of grass.

The background is painted with violet tones, combing the canvas in the direction of the blades of grass.

The brightest blades are painted last.

Observing Nature

In order to paint the textures of any element that comprises a landscape, we recommend that you observe all aspects of nature without limiting yourself to the most idyllic parts . . . It is possible to find grassy terrains in almost any part of a city or town.

Superimposing with Various Layers

Short grass looks somewhat akin to velvet. Even though this type of texture is not difficult to paint, it is necessary to bear in mind that tonal variations can take place on this type of subject matter due to the effect of light.

It is best to start with a textured base of greenish and yellowish tones so that the definitive strokes blend easily into the whole. The aim of this base is to lend chromatic unity to the different applications of color and tones that may result at a later stage in the process.

MORE INFORMATION

• Land textures with oil **p. 58**

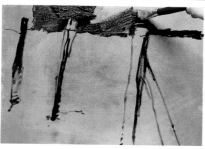

The canvas has been covered with this greenish-yellow color in order to provide it with a stable base over which the texture can be added.

Superimposed layers are applied with very directional strokes using, if possible, a very worn brush loaded with little paint.

Superimposed Layers for Painting a Lawn

Short grass, like the kind in a lawn, is normally painted on a previously colored base. The texture is obtained using uniform brushwork. Furthermore, the brushstrokes must all have the same slant in order to convey the robustness of short grass. A worn hog's hair brush or something similar best produces the effect required for painting a lawn and other examples of short grass. For this reason, old brushes should be kept. After loading the paint from the palette onto the brush, the excess can be removed by wiping the brush against the side of the palette.

First the darkest forms of the shadows are painted.

A worn brush is used to paint the background.

Dry Grass

Dry grass lacks the typical cool colors of damp grass. In fact, the drier the grass, the warmer its tones. Dry grass is normally painted with earthy colors for the background, and burnt sienna for the darkest shadows. The general color of this type of grass would consist of a series of ocher tones; the color should be applied with a worn brush loaded with little paint. The best technique for obtaining the texture of dry grass is to scrub the surface of the canvas with the brush. The most visible and tallest stems are painted with a finer brush using brighter tones.

The most visible stems are painted with a brush loaded with brighter tones.

TECHNIQUE AND PRACTICE

THE TEXTURE OF FLESH WITH WATERCOLOR (1)

Flesh is one of the most difficult textures for the artist to capture, especially when painting with watercolor. The watercolor medium is complex because it entails beginning with the brightest parts of the picture and then adding the necessary washes to model and create the texture.
The foundation of a watercolor painting is the drawing, which should be clearly drawn to indicate the highlights and shadows. Soft and delicate types of skin look smoother than weather-beaten skin, which requires more pronounced highlights.

The preliminary color is almost transparent.

The Preliminary Tones, from Light to Dark

Unlike other media, watercolor requires the painter to work from light to dark. The reason for this is simple: it is not possible to paint a light color over a darker one. Regardless of what the tone or color you choose is, once you apply it over a darker one, it will darken the underlying tone instead of lightening it.

All flesh colors must be harmonized with the surrounding color. They should begin with several pale tonalities from the corresponding chromatic range.

These preliminary layers of color form the color foundation over which subsequent layers of color will be applied. Highlights will be created in those areas that have been reserved.

Once the first glaze is dry, we can add the second tonality that separates the zones in light.

Drying Time

The drying time is one of the most important aspects of the watercolor medium.

If you want to create a gradation, the paint has to be wet in order to prevent undesired breaks from forming. If, on the other hand, you want to reserve several zones, the base color should be left to dry before working with a new color, otherwise you could create unmanageable color blends.

Before the color of the shadow has dried, certain points like the knuckles or the forearm are painted.

The Texture of Grass
The Texture of Flesh with Watercolor (1)
The Texture of Flesh with Watercolor (2)

55

Delicate Skin

Delicate skin should not contain sharp contrasts on its surface. The modeling of the forms is very important, and although variations in tone due to the interplay of light may exist, anything more than a slight tonal change will be too much.

When painting delicate skin you should start by applying a very transparent layer of color. Then, while this first layer is still wet, and so as not to produce any undesired breaks, you can paint a second layer. This will help separate the two main zones of light.

If you wish to heighten the color of a specific zone, before the paint has dried you can apply tiny amounts of other colors to give the skin a slight blush in the parts where it gathers, such as the knuckles.

Weather-Beaten Skin

The texture of weather-beaten and tanned skin is pockmarked with sharp contrasts. Thus, the highlights on the surface of the skin are heightened by the contrast of many shadows. To paint this kind of texture, the artist must begin with a light tone over which he can paint layers of darker colors.

The luminous and almost transparent nature of the first wash permits the highlights to acquire different values according to where they are situated. Last, a dark tone will create denser shadows.

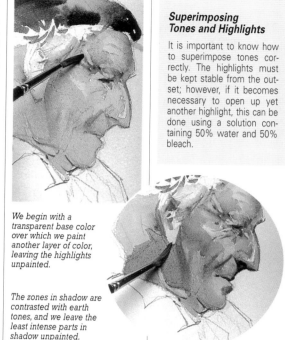

We begin with a transparent base color over which we paint another layer of color, leaving the highlights unpainted.

The zones in shadow are contrasted with earth tones, and we leave the least intense parts in shadow unpainted.

Superimposing Tones and Highlights

It is important to know how to superimpose tones correctly. The highlights must be kept stable from the outset; however, if it becomes necessary to open up yet another highlight, this can be done using a solution containing 50% water and 50% bleach.

The final contrasts are painted with precise strokes over a completely dry base.

MORE INFORMATION

· Pictorial media: Watercolor **p. 28**
· Materials for textures in watercolor: Salt, turpentine, bleach **p. 42**
· The texture of flesh with watercolor (2) **p. 56**

THE TEXTURE OF FLESH WITH WATERCOLOR (2)

Men's skin is different from women's. The textures differ both in color and in softness. Painting different subjects helps the artist acquire the necessary sensitivity for capturing either texture. With practice, the beginner can learn to clearly differentiate between the modeling of female and male skin. Watercolor is a medium that employs many techniques, all of which can be used to create any kind of effect on skin. Masking fluid allows us to reserve those tiny areas that will represent the highlights or droplets of water on wet skin.

Strokes that blend with the background are used to suggest body hair.

The same color used for the chest hair is blended with the background, thus creating a luxuriant texture.

Texture of a Male Torso

One of the main characteristics of the male torso is its hair. To paint this texture with watercolor, we should begin the appropriate way: first, the more luminous tones are painted over the drawing; next, before these tones have dried, the shadows are modeled with darker tones; the brush is then used to spread the paint and soak up any excess liquid. We apply a few loose brushstrokes of dark tones in the chest area to create subtle contrasts with the flesh color. Once the composition is dry, the zones of contrast are heightened with raw umber, making sure no abrupt breaks are produced between the layers of colors. With the same tone that was used for the shadows, and with a very fine brush, the texture of the hair is added. It is best to do this while the base is still slightly wet in order to glaze the chest.

Once the area of the droplets has been covered with masking fluid, the colors are applied and the fluid is then removed.

Texture of a Female Torso

The female torso lacks hair. Its rounded forms are softer than the shapes of the man's body, and the musculature is generally less pronounced. The surface of female skin is soft and does not have abrupt cutoffs.

When painting a woman's skin with watercolor, it is essential to take maximum advantage of the medium's many possibilities. For instance, when painting a zone like the shoulder or the breasts, you can open up light areas by absorbing the paint with a clean, dry brush. This process allows the artist to model the forms softly and create the appearance of a smooth and luminous texture.

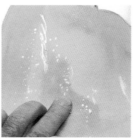

The Use of Masking Fluid for Highlights

Among the many resources available for watercolor painting, masking fluid is one of the most useful.

In this case, we are going to use masking fluid to create the effect of droplets of water on a female torso. Before we begin to apply color, the droplets of water and the path they leave are reserved with masking fluid. Once the masking fluid is completely dry, the entire area of the breast and the arm is painted.

Each one of the droplets is painted with a diluted wash.

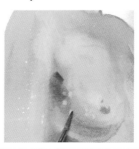

The Texture of Flesh with Watercolor (1)
The Texture of Flesh with Watercolor (2)
Land Textures with Oil

57

The forms are then modeled by increasing the contrast in the dark zones and lightening the lighter zones by absorbing any excess paint. Once the paint is dry, the masking fluid is removed with a finger. Then the droplets of water are painted with a skin-colored wash. Finally, all that remains is to give contrast to each one of the shadows of the droplets.

The Texture of Wet Skin

Wet skin takes on a special shine, especially in the zones that capture light directly. When using watercolor, the lightest areas are represented by the color of the paper itself. To paint the texture of wet skin, it is essential to mark out the areas of maximum light in your preliminary drawing. The next step involves painting the

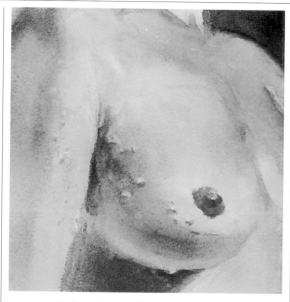

The drops of water are brought out by painting their shadows.

zones in shadow with sienna, without invading the areas reserved for the highlights. The color should be softly gradated until the tones of the lightest areas of the skin have been achieved.

Over the same color base, we increase the contrasts until the half-tones have been created.

The Color of Flesh

Human flesh is not always the same color: it varies according to the light and the surrounding colors. Skin can be painted with green, red, or blue if that is the predominating color of the environment within which the subject is situated.

The contrasts that give shape to the body are increased.

MORE INFORMATION

• Pictorial media: Watercolor **p. 28**
• Materials for textures in watercolor: Salt, turpentine, bleach **p. 42**
• The texture of flesh with watercolor (1) **p. 54**

A final increase in the contrast of the dark areas helps to bring out the highlights of the half-tones.

TECHNIQUE AND PRACTICE

LAND TEXTURES WITH OIL

The texture of earth is so rich and varied that we can only examine a few examples here. This chapter will explain two possible ways of painting mountainous terrain through color and texture. The texture of the ground depends to a certain extent on how damp it is, the type of vegetation found there and, naturally, the light. By using oil we can continually touch up the painting until we obtain the desired texture.

The foundation of a land texture comprises a host of tonal variations. It begins with the application of different tones, greens in this case.

The Variety of Tones

Land can take on a wide variety of tones and colors. Its orography changes in tone and color according to the height and the direction of the light.

Observed from a certain distance, land is rich in tones and shadows. Painting land is never a question of seeing it as a uniform color, for any strip of terrain contains countless tonal variations that depend on distance and height, in addition to the host of tones and colors of the plant-life that exist in such places.

Superimposing Layers of Color

Once the initial wash of color has been painted, the textures and forms of the land take shape from the direction of the brushstrokes used to construct them. The short grass, in this case, is painted vertically. The added color is blended with the first layers. Small patches of bare land, which for obvious reasons are not given the same texture of grass, are painted as uniform patches of color. The different zones of the terrain acquire new values as more colors are painted over the first ones. A series of tiny orange strokes helps balance the definitive texture of the land.

| **MORE INFORMATION** |

- Pictorial media: Oil **p. 22**
- The texture of grass **p. 52**

Tiny vertical strokes to represent the grass are applied over the initial wash of color.

Small luminous patches represent the flowers.

The finish is carried out by adding the color of the flowers over the terrain.

Relief

The foreground is painted with a great variety of tones, each one of which takes up a particular zone in order to define the shape of the land.

With the colors used in the initial texture, each one of these zones is painted so as to integrate them according to the planes they occupy in the composition, while ensuring that they maintain their shape.

Last, a dark tone, such as burnt umber, is used to outline the zones of shadow produced by the stones and rocks.

Painting the Terrain

The enormous differences in the surface of the terrain is conveyed in the painting by means of interplaying contrasts. The texture of a landscape tends toward

First the background is painted with blues to indicate the distance.

blue as it recedes into the distance. Therefore it is necessary to vary the color of the planes according to the distance between them and the observer. The background will contain dark blue colors while the foreground will have the brightest tones, ranging between ochers and greens.

The various tones of ocher and green serve as a foundation on which the texture can be elaborated.

Continuous Practice

The best way to begin painting land is by drawing sketches from nature. This allows you to study a landscape, and see how its different parts can be interpreted by means of different kinds of brushwork.

The colors painted over the base give rise to the form and texture of the landscape.

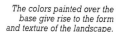

DRAWING TEXTURES

Drawing is based on a linear structure or a series of monochrome washes that together describe a subject, whether real or imaginary. Drawing allows us to examine any sort of texture in its simplest form. Basic drawing media provide us with a wide variety of resources for executing all manner of textures, since they can be done using the contrast between the gray tone and the color of the paper. The final result of a drawing can be further enriched by adding touches of white or of a single color.

A finger can be used to spread gray zones in order to create textures.

The eraser can be used to open up highlights over a blended surface.

Blending with the Fingers and the Rubber Eraser

Basic drawing media, such as a graphite pencil, charcoal, chalk, and conté, although very different from each other, can be perfectly combined together. The one characteristic dry drawing media share is their capacity for blending and the possibility of correcting with an eraser. Other media cannot be erased or blended, for example, for they are more akin to painting than drawing. Dry drawing media can be blended before they are applied by staining a finger with lead and then spreading it on the paper; this can also be done by drawing a line on the paper with the medium and then smearing it over the surface. Later, once the blending has been finished, highlights can be opened up using an eraser.

Blending Lines

In order to create textures in a drawing, such as hair or skin, the artist has to combine different techniques with one medium.

By combining different strokes we can draw hair. This can be carried out over a surface that has been previously blended.

Another texture can be obtained by the blending itself; a controlled blurring technique can be used to create flesh tones in which the highlights are opened up with an eraser.

If you look closely you will notice that the hair was drawn on a blended patch of dark gray.

Building Forms and Their Surface

Any texture of a drawing occupies space in a previously structured surface; this space supports the form with a type of framework. The texture of each one of the model's elements as well as the light is added once the form has taken shape. So when we draw a picture containing different textures, we should work on each particular texture at a time.

Modeling as Texture

One of the most common drawing techniques used for bringing out a determined shape consists of modeling the subject with light and shadow.

The modeling establishes the different zones of light, the highlights and, consequently, the texture of the object being drawn.

Color, Texture, and the Drawing

The drawing does not have to be of a strictly monochrome nature. Among the many possibilities the drawing medium offers, one may find ways of resolving a drawing by means of different color tones, without abandoning the basic concepts of how to draw. The different tones can be used to highlight all manner of textures.

Each one of the parts of this subject has a specific texture that differentiates the materials: skin, hair, fabric, and earthenware.

MORE INFORMATION

- Drawing: Techniques using ink; lines **p. 18**
- Drawing: Techniques using ink, hatching drawn with a pen and reed **p. 20**

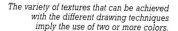

The textures created by light can be understood by studying a piece drawn in chiaroscuro.

The variety of textures that can be achieved with the different drawing techniques imply the use of two or more colors.

TECHNIQUE AND PRACTICE

THE TEXTURE OF PLASTIC WITH WATERCOLOR ACCORDING TO THE LIGHT

In earlier chapters we concerned ourselves with the representation of flesh textures and terrain. The textures of objects possess characteristics that are determined by the way light falls on the model. But regardless of the illumination, the material that the object is made of lends it its own particular qualities. Plastic is one of the most difficult materials to paint, since it does not always have the same appearance. Good examples can be found by looking at objects like plastic dolls and other toys. Although this is a material that can adopt all manner of textures, its surface is always homogenous, and its shine is unlike that of any other material.

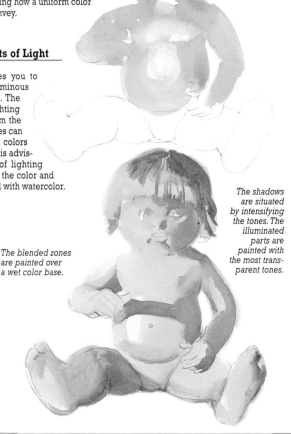

The Color of a Plastic Doll

Plastic can have any kind of texture, depending on the shape, color, and rigidity of the object made from it. We have chosen an everyday object to study the texture and light of this material: a plastic doll exposed to various types of illumination. This will allow us to understand how to capture the form and color of a single object with watercolor.

A smooth texture, which occupies a surface with a variety of planes, is excellent for studying how a uniform color behaves and the textures it can convey.

Results with Different Points of Light

Painting in watercolor requires you to reserve the lightest and most luminous tones and colors from the outset. The study of textures in terms of lighting entails situating the highlights from the first brushstrokes. The darkest zones can be painted initially; however, light colors cannot be painted over them. So it is advisable to practice different ways of lighting the subject in order to understand the color and points of light that can be achieved with watercolor.

The first intonation allows us to evaluate the texture according to the different points of light.

The shadows are situated by intensifying the tones. The illuminated parts are painted with the most transparent tones.

The blended zones are painted over a wet color base.

MORE INFORMATION

• Pictorial media: Watercolor **p. 28**

Modeling Forms

The way forms are modeled depends exclusively on the way light falls on the subject. Although strong lateral lighting creates two tonal zones, soft vertical lighting models the forms of the object being painted by softening its volumes in tenuous gradations.

The least contrasted shadows of plastic are painted over a base color previously dampened, so no matter what color is added, it can be subtly blended into the background.

If we vary the illumination and direct the beam of light from below, the shadows end abruptly. In this case, they are painted over a completely dry color base in order to control the superimposing of layers.

The doll illuminated from above.

Lighting from below produces hard shadows that end abruptly.

Transparent Plastic

When painting the texture of transparent plastic, for example of a container, it is important to paint the color surrounding the object, its zones of maximum luminosity, and the color of its contents. Having painted the background, the interior color is applied. The highlights and lightest zones are left in reserve. Once the color base has dried, the object's darkest zones and the interior forms, fragmented by the reflection of light, are painted. Last, the darkest forms are painted in order to bring out the zones reserved for the highlights.

The Light Has the Final Word

The direction of the light is fundamental for any representation of texture. Direct electric lighting produces a flat texture, whereas reflected or indirect light causes greater contrast and a wider variety of textures.

A procedure for obtaining the texture of transparent plastic.

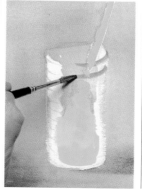

VARIED EFFECTS WITH ACRYLIC (1)

Among all pictorial media, acrylic is the most versatile in terms of the speed with which it can be used and the wide variety of results it can create. The following pages demonstrate the textures that can be obtained with this medium. Such effects are not the product of chance and each one of the textures obtained can be used to paint almost any theme. It is important to remember that a texture is a logical reproduction of an effect, whose sole purpose is to lend a specific appearance to a surface.

Two effects of a wash: one for the background; the other, applied once the background has dried, for the shape of the balloon.

Any form can be painted over a background wash.

Effects of a Wash

There are different ways of painting acrylic washes. This technique is based on tarnishing a layer of recently applied paint while it is still drying.

Acrylic paint dries quickly; in one application of paint we can see how the zone with the least paint dries the fastest. This phenomenon can be exploited to create interesting textures, for example by washing the patches with a fine spray of water to remove the wettest paint.

The technique may appear somewhat complex at first, but with practice these washes can be handled quite well: if you want to allow the underlying color to blush through in certain zones, you must apply more paint to the

area in question. The extra paint increases the drying time of the paint, thus allowing you to remove it at the right moment.

Matter textures obtained with marble dust and acrylic paint.

Matter Textures

Acrylic allows you to execute matter painting very easily.

Any type of material that can be attached to the surface can be added to an acrylic painting. Many types of materials can be used, such as sand, pieces of

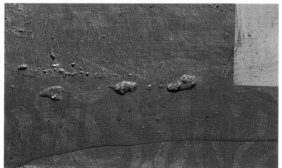

cardboard, or polystyrene (which is very light and can be crumbled into smaller pieces).

When you apply the material to the canvas, the texture becomes part of the painted surface, once the painting has dried.

A layer of background color has been applied and then left to dry.

Blending Tones

All kinds of textures can be obtained with the acrylic medium. One of the richest plastic results is the creation of gradations, since they allow the artist to render highly realistic tonal hues. The texture of clouds or skies in acrylic can be obtained by blending different tones over wet paint.

Example of the texture of clouds painted by blending tones.

Applying Patches of Paint and Scrubbing

Scrubbing the paint can produce a series of interesting textures that allow underlying colors to blush through similar to the way washes do. Scrubbing, however, produces a certain graphic quality, because of the object used to scrub the paint during the drying process. Recently applied paint can be rubbed off with a rag, a piece of paper, or any type of brush. If the implement used is dampened slightly, very clean zones can be opened up to uncover the already dry underlying zones.

The printed texture produces an effect akin to that of a machine.

Printed Textures

Printing is a process based on transferring ink or paint from one surface to another by means of pressure. A rag or a sponge can be used to create a variety of interesting effects which, together with a color background, give rise to highly graphic textures.

Having painted the facial features with black, they are left to dry. Then a layer of white is added on top. Finally, a damp rag is used to remove some of the white to uncover the underlying layer.

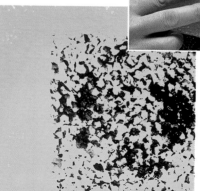

How a sponge is used to create texture on an acrylic ground.

MORE INFORMATION

- Texture and matter painting **p. 8**
- Pictorial media: Acrylic **p. 24**
- Mixing media to obtain textures. Acrylic and oil **p. 34**

VARIED EFFECTS WITH ACRYLIC (2)

The variety of textures that can be achieved with acrylic paint is so wide that it would be impossible to describe all of them. A determined texture that in a sketch book may appear as a simple blob on paper, when situated correctly on the canvas, may denote a shadow, clouds, or even the texture of old wood. Through experimentation, an essential part of painting, you will discover countless other possibilities.

Texture created with a monotype.

The Monotype

The monotype is appropriate for creating graphic prints in a controlled way, while allowing for very interesting textures. To produce a monotype you will need an original image (which is the surface on which you have painted the form you wish to print). The amount of paint used depends on what you want to do on the canvas. The original image is painted using either a brush or a palette knife and placed on the pictorial surface while still wet. Then, by pressing the painted paper or canvas against the blank surface and then removing it, the image of the monotype will be transferred.

Blending Colors

Colors or tones can be blended by caressing the layers of still wet paint and mixing them.

The paint should not be too diluted when blending colors to obtain a sfumato, instead it should have body and be applied in small amounts.

Before superimposing colors, the surface is painted with a layer

First a layer of uniform color is added; then several small tones of color for blending.

Having added the necessary tones, a brush is used to caress the surface of the painting in order to model the forms.

of uniform color. The artist then caresses the still-wet surface of the consecutive layers, executing movements that model the forms.

Sparse or Dense Texture

All manner of textures can be obtained with the acrylic medium, either with dense paint or transparent paint.

A "liquid" texture can be created by mixing very little paint with transparent, acrylic medium. The mix is carried out on the palette until the paint is as fluid as is needed.

Water can be used to obtain such transparencies as well, but we don't recommend it, since it easily destroys the binding properties of paint.

To produce a dense texture you can use any one of the thickeners on the market designed for this purpose. The paint acquires body and can be applied to the

Acrylic texture obtained with a thickener.

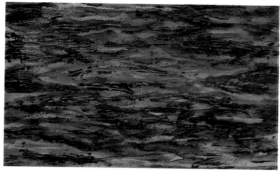

canvas with a palette knife or even as impasto using a brush.

Superimposing Textures

By superimposing different types of textures we can create

A fluid acyrlic tone.

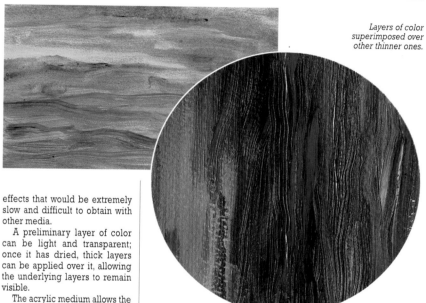

Layers of color superimposed over other thinner ones.

effects that would be extremely slow and difficult to obtain with other media.

A preliminary layer of color can be light and transparent; once it has dried, thick layers can be applied over it, allowing the underlying layers to remain visible.

The acrylic medium allows the artist to superimpose textures of considerable thickness and great transparency over mixed layers of color. This technique cannot be done with any other medium.

Values and Washes

The combination of different types of textures, when applied correctly, facilitates the creation of a tonal values; this can be treated in the same way as other acrylic techniques.

In this example the artist painted the fruit by representing the different values, and then submitted the canvas to a wash process before it had completely dried.

Example of a combination of acrylic textures.

MORE INFORMATION

· Texture and matter painting **p. 8**
· Mixing media to obtain textures. Acrylic and oil **p. 34**
· Varied effects with acrylic (1) **p. 64**

HIGHLIGHTS WITH OIL

Oil allows a great degree of perfection in its finishes, regardless of how complex they may be. The most problematic textures that an amateur painter is confronted with are those of highlights, and transparent or extremely smooth surfaces. Due to the lack of space we are not able to describe each of these challenges and the procedures for executing them, but we are certain that a few examples will be enough to give you a good idea of how highlights and transparencies are painted on the different objects that you may choose to paint.

The highlights of this bottle have been resolved with direct applications of white.

Note how the texture of this vitrified ceramic vessel has been achieved; the secret to its realism lies in the highlights.

Modeling Tones and Highlights

Oil allows a very subtle modeling of forms. The masses can be brought out according to the light that reaches the model, in other words, by means of evaluating the tones.

The many textures on an object containing various surfaces may require the application of direct patches of bright colors so as to lend a suitable appearance to the material that is being painted.

A color that is brighter than the one used to model the object can easily render a shine. In such cases, the texture will be represented with one or several of the highlights reflected off the surface of the object.

A Monochrome Painting and the Study of Texture

An effective way of learning how to paint the highlights of an object is to try obtaining them in a monochrome sketch.

With only two colors, it is possible to intuitively elaborate a surface filled with reflections and shines. For instance, the zones of maximum light on a glass vase will be represented by the white of the paper or canvas. These will be mere strips that follow the plane of the vase. The zones adjacent to these shines will be painted with black.

Applying Highlights to Metallic Objects

Highlights on metal, just like those painted on glass surfaces, must follow the plane of the object on which they are situated. In this case, though, we must include the surrounding colors instead of those behind it.

The variations of color that are produced on polished metal display capricious forms of a highly

One way of learning how to paint the texture of glass is by painting it in monochrome.

graphic nature, which suggest tiny tonal variations in the color of metal (yellow or blue), black, and the colors surrounding the object.

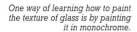

MORE INFORMATION

· Pictorial media: Oil **p. 22**

Metallic highlights make use of black, white, metallic colors, and the colors that surround the object.

Palette Work

You need to develop a good work method for highlights and reflections. It is fundamental for the painter to know how to develop the results of the previous color studies on the canvas. The more hues that can be created on the palette, the greater the choice when it comes to painting the highlights of different objects and painted materials.

The Texture of Pottery

Pottery has different types of highlights according to its shape and the texture of the object being painted, since pottery is not always shiny. Depending on the surrounding colors, highlights will take on a specific hue. A porcelain jug is very smooth and therefore its darkest zones are not so intense; these will in fact reflect part of the light that falls on the object. Porcelain is not normally painted completely white either: the white should always be the brightest highlight of the object; the rest of the colors that form the texture can be whitish, yellowish, or bluish.

How the texture of porcelain objects is captured in a painting.

TECHNIQUE AND PRACTICE

HIGHLIGHTS ON FRUIT

Fruit is a common element in many still life paintings. The technique for imitating the texture of different fruit is entirely dependent on the surrounding colors. Each fruit has its own particular shine, its own color, and its own special texture, therefore the texture of each variety of fruit must be correctly represented, regardless of the pictorial style that is being developed in the painting. It is essential to bear in mind that highlights define the texture of an object, so it is not the same to paint a banana as it is a shiny apple or the skin of a lemon.

This sketch is painted with a color close to the range that is to be used, for instance, burnt umber.

The general color of each one of the zones of the painting are painted, leaving the highlights unpainted.

Golden yellow with Naples yellow and a touch of orange for the lit zone and burnt umber for the shadows.

Highlights in Colorist Works

Colorism is not based on the evaluation of forms, something we have been studying until now. The colorist approach to fruit textures consists of developing the different tones of light through different colors, without a gradation from light to dark.

To block in the color of tomatoes, we should begin with a flat green tone. The zone in shadow is painted with a brown tone, obtained from a mixture of green and red. We begin the other tomato by painting the brightest zone with red and the zone in shadow with carmine. Using this as a color base, we can then add darker, pure colors for the zones in shadow; the highlights are painted with brighter, pure colors.

The Texture of Shiny Skin

Each type of fruit has its own textural characteristics, not all of which are as shiny as apples or grapes.

Some fruit, like lemons or bananas, have very dull textures that don't reflect light; instead, their surface becomes darker or lighter depending on the presence of light or shadow. The colors of banana skin can be developed once an earth color, such as burnt umber, has been used to block in the main shape.

It is important to indicate on the paper the zones of light and shadow. Having evaluated the different planes of the object, we add golden yellow with Naples yellow and a note of orange for the lightest zone, and burnt umber for the zones in shadow.

The colors of the highlights are formed of pure colors that are brighter than those of the initial base.

The first layers of color are flat and pure.

MORE INFORMATION

- Pictorial media: Oil **p. 22**
- Pictorial media: Crayons **p. 30**
- Highlights with Oil **p. 68**

Applying Textures with Crayons

Painting the textures of fruit with crayons is a process that begins with the superimposing of layers of color, each applied with soft strokes, until they have covered the entire surface of the fruit. It is important that the pores of the paper remain open while the colors are being superimposed.

Once you have established the different tonalities of the fruit, special attention must be given to the blending of the colors and highlights so that they accurately convey the surface texture of the material in question. For instance, when you want to paint a shiny apple, the highlights must be large and bright. Other fruit, such as oranges, have a duller, matte shine.

Modeling Forms and Textures

When you paint fruit or vegetable themes, it is especially important to correctly draw their shapes; it is equally important though, to carry out an in-depth study of each piece of fruit in order to synthesize its characteristics on paper. A tomato is not simply red, it has other characteristics that the painter must capture to show the observer what it really is: a round vegetable covered in a smooth, shiny skin, displaying green tones when it is not entirely mature...

The texture of fruit painted with crayons must begin with subtle strokes of the darkest color.

The highlights of the various fruit differ from one another according to the luminosity of their skin.

Choosing the Medium

The artist must know what medium is most appropriate for the work he or she wishes to paint. A mastery of the medium is essential for creating texture and for teaching the artist to recognize each one of the effects that can be achieved on the canvas. For this reason, many artists concentrate on developing their art with one particular medium.

TECHNIQUE AND PRACTICE

CLOUDY SKIES WITH PASTEL

Pastel is one of the most intuitive media, because the way it is applied bears a great resemblance to drawing, thus providing the amateur artist with a great degree of control over the textural effects it can create. Pastel allows subtle gradations and blending using the fingers, an enormous advantage for painting cloudy skies. Over a blended texture that is later fixed, it is possible to apply more colors in order to create clouds of a highly realistic nature.

Several patches of color are painted over the initial blue base and then blended together.

Superimposing Layers

Among the applications of pastel, we should emphasize its opacity and sensitivity in the execution of color blends.

Pastel is ideal for painting skies, especially those that require special attention in the development of clouds. Given that this is a completely opaque medium, the colors and light tones can be applied over any color, even if it is darker.

This is how we begin to paint cloudy skies: Having painted the first layer of blue, we add new bluish tones for the small storm clouds, and patches of white for the more disperse clouds; a slight blending of the layers will make the clouds look more natural; last, some ocher and orange tones are blended in at the lowest part of the sky to indicate the time of day.

A detail of how the different layers of pastel have been painted and blended to obtain the texture of the sky.

A wide variety of gray tones are used to paint clouds.

MORE INFORMATION

· Pictorial media: Pastel **p. 26**
· Mixing media to obtain textures. Pastel and other media **p. 36**

The Use of Gray Tones and their Blending

While in other media the color gray can be obtained by mixing black and white together, mixtures cannot be made with pastel. For this reason, pastel manufacturers have to produce all the tones the artist may require. It may be necessary to use a wide variety of gray tones to obtain the textures of clouds with pastel. You can, however, add touches of blue or even light colors like ivory or white. Over the initial layer of blue, you can apply the first patches of light colors that will serve as a base for the gray tones. The shapes of the clouds can be modeled using the fingers; however, the artist must remember not to abuse this technique.

Monochrome and Light

The wide variety of different pastel tones means that highly detailed textures can be obtained in monochrome paintings. White, in this case, is used to unite the colors. When painting dense clouds, white is used to indicate the main areas of light and separate the different layers of clouds, some of which may be much darker than others. The use of white should be limited because otherwise it may reduce the luminosity and freshness of the final product.

Effects of Light

A pastel stick is made almost entirely of pure pigment. This medium is very luminous when applied on paper, but its freshness and light will diminish the more the color is blended. Having created the texture of the clouds, it is advisable to add several touches of color that will remain intact in order to compensate for the loss of shine produced by the blending that's been carried out. These details may be tiny highlights or atmospheric effects.

Oil Techniques

When you are painting different sky textures, oil can also be used to render interesting effects, such as in the example shown below. The rain was painted relatively simply: after having painted the clouds in oil, the entire background was scratched with the tip of the brush's handle, thus creating this interesting storm effect.

Note how these storm clouds have been painted. Each tone represents one particular pastel stick.

The lightning has been painted with lines of pure color that stand out against the blended background.

The rain was painted by scratching the recently applied paint with the tip of the brush's handle.

TECHNIQUE AND PRACTICE

CLEAR SKIES

The color of the sky is not always blue and less so when it is painted: it can take on leaden or warm tones, depending on the season or time of day; sometimes these tonal variations are used to give the scene a touch of drama. Even a completely clear sky may vary in texture and take on tones that heighten the overall color of the composition and indicate the amount of light present in the atmosphere.

The Sky at Dusk

The sky at dusk and the sky at dawn have similar characteristics. In both cases the color gradations vary within the space taken up by the sky, thus creating different textures of varied color. A common way to obtain the texture of sky is by gradating and blending colors together. Oil and acrylic are the best media for achieving the right kinds of color blends, despite certain differences such as the drying time or the way glazes are applied.

Hues on a Single Color

A clear sky does not necessarily have to be painted with a flat color. It may contain many hues and tones that enrich its texture and help lend it the characteristics of the season or time of the day when it is painted.

Whether you paint with oil or with acrylic, the gradation technique allows you to add different tonalities and colors to a surface that has very subtle changes.

When painting a clear sky, it is important to bear in mind that a gradated primary color applied over another primary will produce a secondary color. For instance, when painting a sky at dusk that contains a gradation from blue to orange, one must avoid blending the colors where they meet, otherwise the result will be green.

Take care when gradating tones not to dirty them.

White is an excellent aid for creating luminous textures on a blue sky.

MORE INFORMATION

· Mixing media to obtain textures. Acrylic and oil **p. 34**

A Leaden Sky with Acrylic

A leaden sky lacks the light of a clear sky. A leaden sky does not necessarily have to be overcast, it may be covered in a veil of mist that makes the color turn opaque, and thus reduce the brilliance of a clear sky. To paint the texture of this type of sky, we must begin with a soft gradation of blues and grays. Having given the colors a few minutes to dry, a very transparent ocher glaze of acrylic is applied over the entire surface. Before it begins to dry, a rag is used to wipe away any excess paint, leaving an ocher veil that will create a strange and dense type of light.

Texture at Dusk with Watercolor

The watercolor medium is completely different from oil or acrylic media. White in watercolor is obtained from the color of the paper itself, and the brightest colors are represented by pale washes. First we paint a yellow wash all over the zone of the sky, leaving the space for the sun in reserve. On the wet background, we paint a bluish gradation, from top to bottom, that will become a green tone. The border of the sun is enriched with orange and ocher tones. Last, we paint a subtle glaze over the dry background to dull the excessively bright sun.

A Variety of Possibilities

Because light is a highly subjective element when interpreting a composition, the artist can develop his artistic creativity while painting the texture of sky. To do this, all manner of techniques must be employed, such as washes, superimposed glazes, etc.

Having painted the base with blue and gray, we leave it to dry and apply an ocher glaze with acrylic.

The process of painting the texture of a sky at dusk with watercolor.

THE TEXTURE OF WATER WITH OIL OR ACRYLIC

Due to its ever-changing surface, subject to light and atmospheric elements,
water does not have a determined texture.
Its capacity to change is precisely what must be conveyed when painting water.
The texture of the surface of the sea or a river at night differs greatly from that of a
choppy sea or a rapidly flowing river.
The reflections and contrasts of water in movement have one particular type of texture
that is achieved by means of a commonly used technique in oil and in acrylic.

From the General Outline to the Details

The texture of water takes on a host of tones and colors. Given that the oil and acrylic media allow for the application of transparencies and opaque layers of color, a dark color is ideal as a base for subsequent additions of much more luminous colors, through which the underlying color will be allowed to show.

The water is painted initially with ultramarine and cobalt blue followed by strokes of light, cool colors that are blended with the background.

Gradually, the entire surface is covered with sweeping, superimposed strokes that blend together in certain areas and are more precise in the highlights.

The entire background is painted in dark blue.

Detail of Repeated Texture

The texture of the different elements is nothing less than the deliberate repetition of an area or a brushstroke. This is essential when painting water, for it is the repetition of brushstrokes that gives water its surface.

Depending on how the brush is used, you can paint a calm sea, a lake, or the rapids of a river, or a storm.

Monochrome Textures

When painting the texture of a storm it is essential to paint the light on the water. Water reflects all light that hits it. A strong storm is generally accompanied by storm clouds, which reduce the light considerably. As a result, the entire scene becomes monochromatic.

The violent movement of water can be interpreted in the painting

The different lighter brushstrokes lend the water texture.

Clear Skies
The Texture of Water with Oil or Acrylic
The Texture of Water with Watercolor

77

with a sharply contrasted texture: very luminous tones that can be perfectly white, and other very dark tones that define part of the shadow of the waves.

The preliminary sketches are used to differentiate the tonal zones, which, as we gradually advance, take on variations in light.

The dark zones have been painted while the parts that represent the light have been left white.

Medium tones and pure whites highlight the darkest zones further and lend them volume.

Chromatic Variety in the Textures of the Highlights

Reflections on water do not have a specific color, but acquire the tone and color of whatever produces them.

As we saw earlier, the color of the water itself is the result of the surrounding colors. For instance, night time reflections on the water can be painted in a concise way: the water becomes a large zone of ocher tones, highlighted by means of horizontal brush-work.

Over this layer of color, the darker reflections are also painted with horizontal strokes. Last, the most luminous reflections are painted with different tones of the chromatic range being used.

Brushstrokes of pure white produce a luminous counterpoint on the crests of the waves.

MORE INFORMATION

- Pictorial media: Oil **p. 22**
- Varied effects with acrylic (1) **p. 64**
- Varied effects with acrylic (2) **p. 66**

Horizontal Texture

If we observe most paintings that include water, specifically as a lake or a sea, we will see that the texture always seems to coincide with the horizon line. For this reason both the shadows and reflections are painted using horizontal brushwork.

Night time reflections on the water.

TECHNIQUE AND PRACTICE

THE TEXTURE OF WATER WITH WATERCOLOR

Thanks to its great luminosity, very interesting water textures can be produced with watercolor. The brightest highlights and reflections in this medium depend on the initial tones applied and the way in which the different layers of washes are superimposed. The highlights on the sea or on a lake are left unpainted when using watercolor.

The highlights are the unpainted parts of the paper.

Respecting the Light Tones

Painting water with watercolor is different than doing it with an opaque medium.

When you work with watercolor, instead of painting the highlights, these are obtained by simply leaving certain areas of the paper unpainted.

Watercolor can be applied over previously dampened paper or over dry paper. If we work on dry paper, it is easier to control the colors and brushstrokes.

By alternating these two techniques, you can isolate the highlights and paint the area that is to be covered with more washes.

The highlights on the surface of the water can also be opened up by removing part of the paint with a dry brush.

The Texture of the Paper and the Stroke

The grain of the paper used is one of the main factors to bear in mind when working on different textures, and especially when painting the texture of water.

An almost-dry brush produces a broken line; if the paper has a

Studying Photographs

The sea and rivers are difficult elements to paint since they are in continuous movement. You should consider taking photographs of them in order to study the flow and movement of water and create similar effects in your own works.

clearly defined texture, a dry stroke will leave highlights in its path; these should be exploited for painting water.

The direction of the brushstroke indicates the direction and form of the different highlights. The stroke should always be applied horizontally, both when opening up whites in the painting, as well as for adding subsequent darker layers.

Small horizontal strokes of ocher enrich the highlighted zones.

MORE INFORMATION

- Pictorial media: Watercolor **p. 28**
- The paper, its grain and different processes **p. 40**
- Materials for textures in watercolor: Salt, turpentine, bleach **p. 42**

The dry zone of the large highlight has been combined with the large patch that was painted while the ground was wet.

The Texture of Water with Oil or Acrylic
The Texture of Water with Watercolor
Trees (1)

79

Uniformity of color is what instills tranquility in water.

Watercolor and the Texture of Still Water

Still water is also painted with horizontal strokes. The absence of highlights and the uniformity of the plane it occupies creates a serene calmness.

Just like with other watercolor techniques, the main highlights are marked from the outset; later, the reflections on the surface will be situated with darker tones. These reflections are painted on a dry background and are then blended with a final layer of gradated color.

It is essential to apply the stroke in a horizontal direction when painting water.

The first patches of color are applied around the zones reserved for the highlights and used to paint the shape of the breaking wave in negative. The blending of the wave with the sky is carried out on a wet background.

The Force of a Rough Sea

A rough sea acquires a multitude of highlights. These can be reserved from the very first application of watercolor on paper.

To achieve this texture, we must take great care to combine the zones to be painted on a wet background (such as the crest of the breaking wave, which blends with the gray sky) with those that are to be painted on a dry one. The area that represents the lower part of the foam is a good example.

The stains of color in the lower part of the painting have been painted on a dry ground, indicating the darker zones.

TECHNIQUE AND PRACTICE

TREES (1)

The amateur artist must pay close attention to trees. Each species has its own particular texture. Depending on the leaves and branches, the stroke will be applied either more compactly or more loosely. You can paint with wide strokes (long strokes) or dense strokes, according to the shape of the leaves and type of foliage the tree has.

The structure was painted in a linear way using one color.

Disperse Branches and Leaves

To paint this type of texture it is convenient to master the structure of the tree and its branches. The branches are brought out by painting contrasts between the different highlights.

MORE INFORMATION

- Pictorial media: Oil **p. 22**
- Trees (2) **p. 82**

The shape of the branches is added with a swift stroke.

The lightest linear tones are painted to create light on the branches.

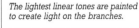

*The form of the shadows
is described with a dark tone.*

*A fine brush is used to paint
the lightest zones of luminous green.*

Evergreen Trees

The texture of certain trees, such as firs and other conifers, require a more gestural stroke for the branches.

The Texture of Bushes

Bushes and shrubs have a dense texture, therefore the painting must start out as a mass of color with strokes that suggest the length of the branches in conjunction with the leaves.

*The highlights
and shadows
are alternated,
bringing out
the texture by
scraping the
brush over
the surface.*

*The first application combines
the two tones of light in a swift stroke.*

*With loose brushwork, and alternating the tones of
light and shadow, the texture of the bush is obtained.*

TREES (2)

Trees respond to light in different ways, partly because of the tonality of their leaves and their texture. The crown of a tree is what gives it its texture. In order to represent the texture of a tree with paint, the artist must use even brushstrokes that clearly define its different elements, including small variations of leaves and branches, while maintaining the unity of the form.

The zones corresponding to the shadows are painted.

The painted form is isolated by painting the background color.

Light on Foliage in the Fall

During the fall, the treetops of certain trees are more compact and golden. The contrast between the zones of light and shadow is very important.

MORE INFORMATION

· Pictorial media: Oil **p. 22**
· Trees (1) **p. 80**

The texture is completed with yellow and ocher.

The Texture of a Palm Tree

The texture of a palm tree requires a structure more akin to a drawing than a painting. Tiny highlights are added to the background color to render the tree's fruit, and long luminous lines depict the highlights on leaves and the dried zones.

The trunk is painted with dark violet; the trees are drawn with loose strokes of dark green.

The texture of the leaves is painted with long, darker brushstrokes.

The Crown of a Pine Tree

Among the many coniferous trees, the pine tree is the one with the greatest density and variety of tones, ranging from ocher to dark green. The brushwork is used to build up the different planes of the crown.

The texture of the leaves has been represented with yellowish touches and lines of ocher.

The structure of the pine tree is painted with a dark tone.

All manner of brushstrokes are used: short ones for the texture of the shadows, concise blobs for the highlights.

TECHNIQUE AND PRACTICE

TEXTURES OF ANIMALS
WITH SMOOTH SKIN

Many animals do not have fur, nonetheless their skin has very different textures.
Some, like the rhinoceros, have thick skin that is as hard as armor; other animals,
such as seals, have a completely smooth and shiny skin, thanks to their
thick coat of short hair suitable for life in the water.

The texture of smooth skin requires a drawing that clearly defines the forms that have to be covered.

A gray tone is the first step towards darkening the forms.

Blocking-In and Patches of Paint

As with any painting, we would not be able to obtain the textures that model the shape of the animal without a good drawing. Animals with smooth skin require a more detailed drawing than long-haired ones, since the hair of the latter conceals their basic shape.

After the detailed blocking-in of forms, we establish the color base and apply darker tones to create contrast.

For an animal like the rhinoceros, whose skin is thick and lacks chromatic connotations, we begin by applying gray to the whole shape (obtained from a mixture of burnt umber toned down with white); then, a grayish version of sienna is painted over to help bring out the texture of the mass.

The texture of the skin is painted without highlights, so that the lightest tones are never pure white.

Texture Obtained Through Variety

Animals that lack fur may have monochromatic skin; however, this does not mean they should all be painted in the same way. Certain special techniques exist for lending the skin a suitable appearance.

Textures such as the skin of this rhinoceros or the seal on the opposite page, can be achieved with a brush; however, they can also be obtained using different materials, such as a sponge, that will print a porous texture.

There is a wide variety of techniques that can be used to give smooth skin the appropriate appearance; for instance, the skin of the rhinoceros, which is completely rough and matte, has no highlights, and therefore the

changes in illumination are insinuated by slightly lighter tones.

The first step is to reserve the highlights.

The Highlights of Smooth Skin

Shiny skin, such as the seal's, contains within its texture a great variety of contrasts that alternate between completely dark zones and medium tones or luminous whites.

In order to obtain a wealth of different tonalities, dark skin should never be painted black, but rather with different dark tones.

When you establish the tones of the skin, the highlights must be indicated. This is carried out in two stages for shiny skin: first the reserved zone is painted over with a tone that is far lighter than the rest; then the pure white zones are situated to complete the texture.

Pure white highlights are added over a tone that is brighter than the rest.

Scales as Texture

When painting the texture of scaly skin, the most important point to bear in mind is that the highlights must be situated all over the surface, alternating them with a variety of blue tones that will lend the whole a metallic quality. The violet and warm hues give the fish its organic aspect.

MORE INFORMATION

• Pictorial media: Oil **p. 22**
• Pictorial media: Watercolor **p. 28**

An example of scaly skin.

The Best Place for Learning About Nature

The artist can go to the source in order to study the textures of different animals by paying a visit to the local zoo. In fact, many painters spend a great deal of time drawing sketches from nature of the animals they wish to paint. This allows the painter to become familiar with the different textures of fur and pelt, as well as the skin of reptiles and birds' feathers.

HUMAN FLESH ACCORDING TO THE LIGHT

The color of human flesh is not always the same, for depending on the way the light falls
on it, it will acquire different tones.
In the studio, skin reflects the light from the spotlights. Furthermore, it reflects the colors
of the surrounding objects. Depending on the amount of light and how
directly it falls on the model, the texture of skin can vary radically,
appearing either soft or hard.

The zones of shadow are painted.

The lightest parts are situated.

*The textures of
the skin incorporate
all the surrounding colors.*

The Texture of Flesh According to the Surroundings

Lighting is one of the main factors that contribute to the texture of any type of skin. When painting human skin, it is useful to remember that its tones are affected by the colors of the place in which the model is situated. Before you start painting the figure, it is worth studying the color of the background. Once the main drawing is finished, the background should be painted, followed by the zones in shadow using a dark

Dark Skin and Artificial Lighting

In addition to melanin, the color of skin depends on the light that reaches its surface. The texture of dark skin is affected by the model's surrounding colors, and by the color of the light shining on it.

Young, brown skin illuminated by artificial light takes on a host of tones. The first ones to deal with

in a painting are the light and dark ones. As the texture is elaborated, the reflections and tones of the surrounding colors are added.

For instance, if you look at these images, you will see that the first areas to be painted were the darkest ones, followed by the lightest ones, and lastly the reflections produced on the skin.

MORE INFORMATION

- Academic painting **p. 6**
- Pictorial media: Oil **p. 22**
- The texture of flesh with watercolor (1) **p. 54**
- The texture of flesh with watercolor (2) **p. 56**

Textures of Animals with Smooth Skin
Human Flesh According to the Light
The Color of Flesh from Nature

87

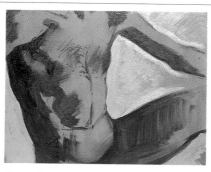

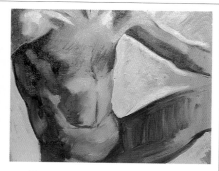

First the figure was drawn and then the background was painted.

When painting the colors of the skin's texture it is essential to bear in mind the model's surroundings.

color. Last, the lightest zones are painted with luminous tones that take into account the surroundings of the figure.

The Color of Pale Skin

The texture of pale, white skin depends exclusively on the colors surrounding the figure. This example shows the process used for painting the skin of Venus in Velázquez's mirror. If you look carefully at the colors being used, you will see that most of them contain a large amount of white. The figure's volume, nonetheless, is not free of shadows. The flesh tones are obtained by studying various mixtures of both light and dark colors on the palette.

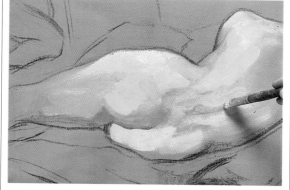

The different color values have been carried out with bright, light tones.

Altered Tones

It is interesting for the artist to practice painting textures with the different colors seen at a first glance on the model. You can paint using only green or blue tones and carefully studying the values of each color added. This is a highly recommendable exercise for studying the relationships between the different colors of a determined palette. At the same time, this is a particularly entertaining exercise that can prove tremendously creative.

The process of lending texture to skin with a varied range of colors.

A Different Medium

Colored pencils also produce very luminous skin textures. This medium lies halfway between drawing and watercolor painting, since in addition to the usual pencil techniques, the transparency of the medium allows texture to be obtained by superimposing layers of color.

The colored pencil medium is based on superimposing layers of color.

THE COLOR OF FLESH FROM NATURE

Depending on its condition and smoothness, skin can take on many textures when exposed to sunlight: the weather-beaten face of an old seaman is not the same as that of a child. The different textures are expressed not only through color, but also by the possible inclusion of wrinkles and highlights. Highlights on skin vary depending on race; for example, dark skin has a characteristic shine and does not redden in the same way that paler skin does.

The first step is to draw the general outline.

The highlights and contrasts are added at the end in specific places, without blending the color with that of the background.

Weather-Beaten Skin

To paint the texture of weather-beaten skin, it is vital to study the main points of contrast on its surface, including the highlights and the zones containing wrinkles and lines.

Once the preliminary sketch has been drawn using any medium, we begin painting the general color, ignoring highlights and contrasts for the moment. This first layer of paint is used to situate main tonal areas: the flesh color of the face is obtained from a mixture of carmine, white, and a touch of sienna or burnt umber. The darkest zones are painted with less carmine and white. Last, tiny notes of blue enrich the zones of contrast, and the highlights of the skin are painted with precise dabs of white.

The Color of Skin and Its Highlights

The texture of colored skin is as varied as the texture of white skin. It boils down to how the light falls on it. The first tones to be applied on the canvas are the dark ones for the shadows. Burnt umber is an appropriate color for this task; one must begin by painting the main shadows without attempting to adjust the tones. Once the shadows are in place, the rest of the texture is painted with the same color lightened with white. Although the skin tones can be warm (slightly orangish), the areas with indirect lighting and the half-tones may tend to be cooler (bluish). Finally, the brightest highlights are painted with almost pure white tones.

The colors of the illuminated zones include orange and blue tones.

Human Flesh According to the Light
The Color of Flesh from Nature
Textures of Animal Hair

89

Means and Methods

All pictorial media are suitable for painting textures. The important point is to always follow the procedure correctly. Every medium has its own possibilities and cannot be compared with others; for example, a flesh tone obtained with watercolor has little to do with one painted with pastel. No medium is better than any other, it all boils down to the individual painter's procedure.

Realism Through Pastel

Pastel allows for highly realistic textures, since highlights can be achieved intuitively and by means of successive layers of thoroughly blended colors. By blending with the fingers it is possible to reach a high degree of realism in terms of texture. Highlights with pastel are obtained by applying very bright tones over gradated flesh tones.

A Recreation of Baño del Caballo *(1909), by Sorolla. The colors are pure, but appear perfectly natural within the context of the surrounding colors.*

Texture According to the Light

It is always possible to paint the texture of skin using pure colors. Skin exposed to daylight reflects the color of the surrounding elements, including the atmosphere elaborated by the painter.

When you paint a skin tone with bright colors like pure orange and ocher, the rest of the painting must receive a similar treatment, thus allowing the observer to perceive the skin's texture as natural.

The lightest tones lend the skin its real and luminous aspect.

TECHNIQUE AND PRACTICE

TEXTURES OF ANIMAL HAIR

Animal hair varies in length and thickness according to the species of the animal. The texture of the hair of different animals can be painted using various techniques: scrubbing the brush against the canvas, applying short brushstrokes, or even creating tonal blends. Each one of the different kinds of hair can be expressed through a specific painting technique.

The main lines of the structure have been drawn.

The Texture of a Dog's Shaggy Hair

Animal hair always maintains its shape, because it covers a well-defined structure. Frequently, the novice makes the mistake of drawing the animal with the hair included, a practice that leads to a deformed mass. The reason this happens is because the artist does not begin with a drawing that defines the animal's internal structure. The drawing is indispensable for achieving the texture, since it provides the exterior coat with a framework on which to sit. A preliminary sketch need not be complex, but it must be complete so that regardless of what texture is to be added, the result will look like the animal being represented. In this case, the internal structure of this shaggy dog is drawn before attempting to define his coat.

MORE INFORMATION

· Pictorial media: Watercolor p. 28
· Additives for creating textures p. 32

The different zones of light and shadow are now gradually applied over the basic structure.

Once the preliminary color has been established, the texture is enriched with strokes of pastel and brushstrokes of gouache.

Finish with Mixed Media

Texture is painted so as to "dress" the animal's internal structure. The use of charcoal and pastel allows us to quickly block-in the broad masses of the dog's face. The light areas correspond to the highlights and the dark ones to the places where the hair begins to darken. The initial patches of color must be exact and the part of the dog being painted must be modeled by the direction of the brushstrokes. Once the entire head has been painted with a combination of blended and fresh color, the remaining textures are created using luminous strokes that allow the background to show through. In addition, layers of tempera are applied and then scratched with the tip of the brush handle to create the finer hair.

Observe and Synthesize

Observation is the artist's most important tool for capturing and representing a certain texture. To do this, it is essential to draw a sketch. One of the best media for executing textures in a sketch is pastel, since it facilitates graphic superimposing over blended areas.

The Texture of a Bear's Hair

A bear's coat is characterized by longish hair of a dense nature. This gives the animal a soft appearance. On canvas, a bear's coat consists of several alternating layers of brown ocher and orange ocher.

The shape is recreated with the color but without relying too much on the direction of the brushstroke. The shape of the hair should only be conveyed when painting contrasts or where the shadows cut off abruptly at a lighter zone. This is done using a zigzag form. Over a totally defined color base, the hair is painted with a dark brown, especially in those zones where dark and light zones meet.

The darker forms are painted over the original color base to give shape to the hair.

The hair takes shape in the darkest zones.

Short Hair

In short-haired animals the interplay of contrasts on the surface is produced by the texture. The softness of the hair is most prominent in the areas where the different patches of light meet. For instance, if you are painting the hair of a zebra, the dark zones in shadow indicate the length and texture of the hair.

Short hair is painted based on the contrasts produced over the whole shape.

TEXTURES OF REPTILES

Unlike mammals, the skin of frogs and reptiles is smooth, yet it maintains a characteristic shine. The skin of amphibians is shiny but it also displays a certain viscosity when it is dry. In the case of terrestrial reptiles, their skin is characterized by tiny scales which, on certain parts of their body, reflect light with litmus tones.

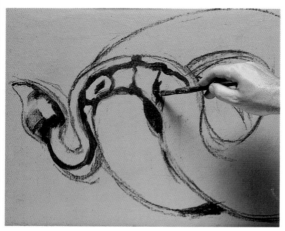

Texture and Drawing

It is recommended for the artist to practice textures by drawing. In fact, many textures are created by means of the interplay of lines drawn on a previously painted surface. These lines may define scales, hair, highlights or veins in wood.

A snake's skin adapts its form according to the position of the body.

The colors are applied while leaving the highlights untouched.

The Snake's Skin

The skin of reptiles is characterized by shiny scales. Snake's scales are tiny and extremely brilliant, a fact that makes the whole acquire a special brilliance and luminosity quite unlike a fish's wet scales.

After correctly defining the animal's anatomy in a drawing, the colors are applied according to specific areas on the skin's surface. The best way to draw the snake's skin depends to a great extent on how it twists and bends, since the forms of the drawing must adapt to the position of the animal, synthesizing the face and concentrating more on the undulating body.

Highlights

The snake's scaly body and that of other reptiles makes the light reflect off the surface and display certain colors that are a mixture of the light and the color of the scales themselves. To paint this texture it is important to combine the tones of the reflections with suitable brushwork. Using a fine, flat hog's hair brush, the luminous zones are painted to show where light is reflected off the skin. These textures are created with hatched-like brushwork, crisscrossing lines that, from a distance, create the illusion of highlights on the scales.

The highlights of the skin are applied in the form of crisscrossed lines over the initial color base.

Chromatic Variety

In general, reptiles present a wide variety of colors. Certain lizards, such as the iguana, have various completely different textural zones over their bodies; the scales on their head produce remarkable reflections while their body is more matte in nature.

The shape of the animal is drawn against a dark background, in order to bring out the luminosity of its colors. The contrasts and shadows are determined during the initial color application. Using a color scheme that combines green and earth tones with brighter colors, the highlights are painted and the texture of the skin is drawn.

First the background is painted to create a greater contrast of form.

Scales in Watercolor

Painting the texture of a reptile with watercolor begins with applying a base of several transparent colors, through which the form is modeled by opening up lights areas with a clean dry brush, while the wash is still wet. Once the color base has completely dried, the scales and lines of the animal are painted with a fine brush. The highlights are left open from the outset; the texture of the scales, having been painted with lines, will show through.

The first layers of color serve to define each part of the body and model the foundation of the texture.

MORE INFORMATION

• Pictorial media: Watercolor **p. 28**

The creases and wrinkles are given a linear treatment and are painted with luminous tones that imitate skin.

The general texture is created by means of a series of washes. Once they have dried, the shapes of the scales are drawn. The remaining texture is achieved through a series of crisscrossed lines painted on a dry background.

PLUMAGE TEXTURES

Each species of bird has its own particular type of plumage, not only in terms of color (which is very important for recognizing a bird), but also in terms of softness, smoothness, and shine. Different media offer different techniques for resolving each type of plumage texture; it is not the same to paint feathers with oil as it is with watercolor.

The texture of a rooster's feathers is painted using long brushstrokes and contrasts that indicate the dense volume of the feathers.

The Texture of a Rooster's Feathers

A rooster's feathers combine bright colors (red, earth tones, and dark gray for the contrasts) with long brushwork to indicate the texture of the plumage. The long brushstrokes are emphasized with earth tones, and the main shadows are drawn with fine strokes. Depending on the medium you choose, the superimposing of colored layers will be carried out in different ways. With watercolor, one must wait for the first color to dry before applying another one on top. Furthermore, when painting with watercolor the white areas are represented by the color of the paper, for it is not possible to lighten one color by adding another, as can be done with other pictorial media.

The shape of the feathers is defined with long and contrasting brushstrokes.

The Soft Texture of the Budgerigar

The budgerigar and other domestic birds have a soft texture. They have small feathers except for the ones on their tail and wings. The breast of small birds can be easily resolved with a subtle color gradation; lines or sharp contrasts are not necessary on such a delicate volume. The wings and the most contrasting zones are painted with long brushstrokes, which allow for future touch-ups with shorter strokes and slightly darker colors. If we

The color gradation of the breast and several long brushstrokes applied over the wings are the first steps in defining the color of this painting.

compare the textures of smaller birds with those of larger species, it is easier to understand how the plumage should be treated.

From the Mass of Color to the Brushstroke

To paint any type of texture, start from a surface that has been covered with masses of color, that is, begin with a color over which more detailed work can be added. The detailed work will consist of brushstrokes that become more defined the closer they get to the top layer. In watercolor, the color base must always be more luminous and brighter than the successive ones. In other media it is possible to use a dark color for the textural base and paint lighter and more luminous colors on top.

The way to paint the texture of small birds is similar to how the budgerigar is done, that is, by varying the color.

The feathers of the top layers are dark and are painted with small strokes that follow the direction of the wing.

MORE INFORMATION

- Pictorial media: Oil **p. 22**
- Pictorial media: Watercolor **p. 28**

The brushstrokes are longer in the wings and more contrasted in the breast.

Luminosity and Color

The feathers of large tropical birds are long and narrow. To paint these types of feathers, it is essential to start with a color base, and then apply long strokes that follow the same direction as the feathers themselves. The zone with the most homogenous texture is the breast; its color has fewer variations in tone that any other area where the strokes are evident.

The different chromatic zones are applied with sweeping brush-strokes that cover the zone and give the bird its form.

Practice Techniques

One good way of learning to paint the textures of different kinds of plumage is to try painting them with different media and see what kinds of results you achieve. It is important to always paint a color base before attempting the texture with other colors. The color base should be allowed to show through the subsequent brushwork that is added to build the plumage.

Original title of the book in Spanish: *Texturas*
© Copyright Parramón Ediciones, S.A. 1996—World Rights.
Published by Parramón Ediciones, S.A., Barcelona, Spain.
Author: Parramon's Editorial Team
Illustrations: Parramon's Editorial Team
Copyright of the English edition © 1997 by Barron's
Educational Series, Inc.

All inquiries should be addressed to:
Barron's Educational Series, Inc.
250 Wireless Boulevard
Hauppauge, New York 11788

International Standard Book No. 0-7641-5061-8

Library of Congress Catalog Card No. 97-73853

Printed in Spain
987654321

Note: The titles that appear at the top of the odd-numbered
pages correspond to:

The previous chapter
The current chapter
The following chapter